UNDERSTANDING
PHYSICS TODAY

UNDERSTANDING PHYSICS TODAY

BY

W. H. WATSON

CAMBRIDGE

AT THE UNIVERSITY PRESS

1963

PUBLISHED BY
THE SYNDICS OF THE CAMBRIDGE UNIVERSITY PRESS

Bentley House, 200 Euston Road, London, N.W. 1
American Branch: 32 East 57th Street, New York 22, N.Y.
West African Office: P.O. Box 33, Ibadan, Nigeria

©

CAMBRIDGE UNIVERSITY PRESS

1963

Printed in Great Britain at the University Press, Cambridge
(Brooke Crutchley, University Printer)

CONTENTS

PREFACE

This book is a remote descendant of *On Understanding Physics* written a quarter of a century ago. The new writing was stimulated in part by the reprinting of it, and in part by a period of active philosophical reflexion during the past three years. This activity was sustained by the desire to understand how it is that physics goes on with ever-increasing pace to reveal new structure in the world in which we live, matching the achievements in chemistry and biology and in applied science, yet the theoretical apparatus for managing microphysical phenomena seems to involve us in some philosophical confusion.

We have no difficulty concerning the existence of the cells in living matter or of the molecules composing them. We have scientifically effective descriptions of many processes that play an essential role in the real world. We commit ourselves confidently to new discoveries in genetics, physiology, biophysics, biochemistry, geophysics, and astronomy, as well as physics, so long as they are presented in terms of our classical intuition associated with the use of ordinary language. But when mathematics is used to design theoretical methods for managing microphysical phenomena we are deprived of the supports on which we have been accustomed to rely in our naïve approach to nature, and our treatment of atoms, nuclei and elementary particles seems to present them disembodied in space and time. Are we to regard this as a mystery? Or, are we to say that we still stand in ignorance of how we got into this situation, and how we might extricate ourselves by a better understanding of our metaphysical *naïveté* in the face of representing the world and how we use our representations?

The main bias of my thinking is against misplaced erudition that conceals what should be laid bare and made clear. Unfortunately what is clear for one man may not be so for another, for this

vii

depends on the accidents of experience and the influence of special training. So a book concerned with philosophising about physics is in the first place a personal document. It comes into being in response to difficulties in understanding physics, but it also has metaphysical roots. Only if it makes sense for other men and is relevant to their experience and inclination towards life does it serve a true purpose.

A book dealing with ideas usually depends on serial exposition in a form that corresponds to a progressive elaboration of logical structure, going from the elementary and familiar to the complex and sophisticated. But philosophising does not usually go in this way. It is not the function of philosophy to evolve the formal systems of science. It stands outside them, examining how we are involved with them. Formalising the varied activities of philosophising is effective only for academic formality. It does not last long as an intellectual influence. As evidence of this, examine the formalisations about physics (as opposed to formalisms of physical theory) at the hands of professional philosophers and of mathematicians and scientists who have attempted to formalise what clearly cannot be formalised.

Although he may not acknowledge it, the philosopher often engages in theorising about human behaviour and thinking processes. Unaided by scientific investigation and imagination, this activity yields little insight that is not enshrined in great literature of the past. With the aid of science in our time, however, particularly through the study of neurophysiology and the science of automatically controlled machines and computers, men have laid hold on new ideas for thinking about thinking and behaviour, particularly in learning. The academic philosopher needs to remind himself that the learning process is a continuing part of adult intellectual life. It is not confined to the kindergarten, although there learning is seen at its automatic best, as it is in birds and animals, and modern science is revealing much about it. That does not mean, however, that the new conceptions on which scientists rely to advance understanding of learning and behaviour have any finality about them. In so far as scientists accept a particular point of view, they are merely committing themselves to a practical formalism until something better evolves. And this is the teaching of history about science. It reaches out to the new and unexplored, while consolidating and engineering efficiently its conquests of the past, and divesting itself of the habits acquired in the historical accidents by which it evolved—just as happens with any good invention. One does not want to keep talking about the earliest airplanes as an essential part of aerodynamics or flight engineering. They have a human interest of course, but not an essential intellectual interest in the context of flying and the engineering of flight today.

It is important how we are inclined to judge these matters because this reveals the bias of our minds towards philosophy. I mean the philosophy that attempts to enlighten and get us out of the pathological state of always looking back. If we do occasionally look forward, we are already far behind the front of current intellectual achievement in science. Thus we are restrained from purposeful participation in the life of our time.

The basic quest is for an intuitive grasp of atomic existence—even in the face of authoritative injunctions in treatises on quantum mechanics and field theory not to attempt this. This quest is not to be fobbed off by formalistic explanations, or impeded by grotesque philosophical language and style. The goal is a practical one and is to be attained by keeping the discussion matter-of-fact and avoiding philosophical pretence. The goal is practical because my interest is ultimately to help the learner with philosophical difficulties. Pursuing it requires attention to detail.

The clues to understanding are followed along many tracks in an involved network of lines of connexion which intersect and overpass each other. Understanding consists in making a clear map of these connexions and learning it so well that one can resist automatically the temptation to take the wrong road in thinking. The philosophical questions that arise resemble those that have been generated by philosophers about ordinary language. Whereas in dealing with ordinary language we should refer our questions to the context of its use in the ordinary affairs of life, in physics we should describe what physicists do, and how they use the ideas, how they speak about them in doing physics, and have regard to the fact that physics goes on conquering more territory and achieving technical control of it in spite of philosophy.

The philosophical difficulties about ordinary language arise from formal preconceptions about human behaviour and the details of how we learn and speak and act in using language. These preconceptions could hardly continue to dominate academic thinking were there a proper appreciation that training is necessary for all intellectual achievement. So we require a better conception of how man learns based on understanding man as an animal, instead of following the socially embedded traditional approach to these matters. But such a conception must be based on scientific theory. It will not be static: it will evolve. And each generation must learn anew.

Training and basic human experience supply the sinews for operating with physical reality. They facilitate the unformalised elaborations by which theory is brought into relation with experience in the laboratory, the observatory or in the field. This process does not work only in one direction; there is feedback from that experience. This leads in the long run to new methods of representation, new forms of training, and so on. Thus physical science grows, and we shall not make much progress with philosophical elucidation of intellectual perplexity about physics unless we refer to training and experience, and see what men do with things, with words and with ideas, and note how they behave, just as Wittgenstein did in philosophising about meaning.

What traditional philosophy leaves out is the role of training in intellectual life. This is the essential support of the unformalised elaborations of thinking and acting. It is the necessary practical basis for using symbolism, language and tools.

Nowhere, perhaps, is the foregoing consideration of greater force than in relation to meaning, calculation and the representation of physical nature. The grammarian's approach to these matters just will not do. Meaning is learned through use and this practical experience establishes ability to go on. How it may do so has been speculated on by the active group of scientists who face the difficult practical problems of designing machines to replace men. I do not wish to pursue this point here, but I make it because it is relevant to thinking about the philosophy of science and reflects on the intellectual tradition in terms of which much of the writing that claims to be philosophy of physics is done. Physics is not based on any formal philosophy; it is based on training and practice and on human behaviour that has evolved with the growth of experience in doing physics.

Philosophers write as if there were an alternative to physics to perform its function, or as if physics were unsatisfactory and needed to be supported by a base or stand of some kind that is not physics, or framed by a frame that is not physics, or as if physics were black and white and should be coloured, or as if physics were a corpse and needed to have the breath of philosophy breathed into it in some kind of artificial respiration so that what is 'mere matter' could have a soul.

The soul of physics is given by physicists who think about it, who do experiments, discuss it, write about it and teach it. This is the only kind of soul worth having. The rest is a sort of pathological morbidity that keeps a man from learning about nature, and discourages real participation in that creative process. In good health it is not natural. Philosophy, as Wittgenstein once remarked, ought to liberate us from the idea that there is a kind

xi

of academic doctor who can do things for physicists and other scientists that they are incapable of doing for themselves. Primitive science does not minister to modern intellectual needs in doing scientific work. Since men do not long depend on ineffective means when they have an idea of accomplishing something worth while, they will invoke a renaissance of learning directed to enthusiastic purposeful intelligent living. This renaissance should show first in universities. That it will, in fact, remains to be seen. The forces creating fresh intellectual awareness related to real life are operative now, I am sure. They will cause an accelerated development. This is going on today in scientific research institutes some of which seem to flourish in spite of their separation from the traditional academic environment. I hope that this book is written in the spirit of the changes in outlook that I think I discern.

The first three chapters of this book are intended to familiarise the reader with the style of thinking and attitude to philosophising with which I have approached the basic metaphysical questions concerning existence, substance, motion and atomic connexion that are discussed in the remainder of the book. In the course of these discussions we shall reflect on how the methods of representation in quantum mechanics and quantum field theory may be regarded as connected with metaphysical assumptions. On the basis of these assumptions which are exposed in contexts with which we are familiar in ordinary life, our goal is to look on the mathematical system of quantum mechanics not merely as a mathematical invention that works well in our experience, but as the reasonable necessary way to represent atoms in the continua of classical physics. In the process we shall develop what is the logical essence of the idea of atomicity, not in the traditional Greek sense but in a sense that seems to express the content of our thinking about nature as revealed by modern physics.

In the course of the discussion it is necessary to refer to basic mathematical ideas connected with representation, in order to expose logical form, not to elaborate the mathematical developments by which the effective representations of modern physics are worked out. This book is not intended to explain quantum mechanics and how it is applied. Neither is it concerned with the mathematical systems by which the formal structure of quantum mechanics can be presented. Our interest is metaphysical, examining the roles of classical representation and ordinary language in the investigation of microphysical phenomena, and to put it briefly, concerning ourselves with the natural philosophy of atomic phenomena. For the most part our activity is to exhibit what we do in a new light.

To President Bissell and the Board of Governors of the University of Toronto I am indebted for the leave of absence from the University that made possible the completion of this work and its preparation for publication. I wish to thank also President Sterling, Professor Suppes and Professor Schiff whose hospitality made my stay in Palo Alto and use of facilities at Stanford University so congenial.

W. H. WATSON

Palo Alto
CALIFORNIA
December 1961

xiii

Amber was a favorite material for these experiments, and the Greek name for amber is *elektron*. Thus does language often record history.

ROBERT R. WILSON and RALPH LITTAUER in *Accelerators*

CHAPTER I
PHILOSOPHISING ABOUT PHYSICS

It has become a commonplace to state that modern physics, in so far as it depends on the idea of a physical field, dates from Clerk Maxwell whose electromagnetic theory was vindicated by the remarkable experiments of Hertz. In the introduction to the book *Electric Waves*, in which his papers on this subject were collected, Hertz examined Maxwell's theory in a philosophical way. What influence was exerted in the development of physics by Hertz's writing? It is very hard indeed to ascribe a direct influence to Hertz in spite of the cogency and insight that inspired his writing. Nevertheless, so far as Maxwell's theory is concerned, his view of the matter came at length to prevail. Are we then to say that philosophising about physics is ineffectual?

Hertz's writing was pithy and to the point. It was no mere literary activity furnished with the inadequate interpretations and picturesque allegories that serve today to attach a wondering general public to the professional work of physicists through the popular writings of those whose achievements in science have won them fame. It was directed to matters of substance affecting the attitude of the scientist to his methods of representing nature, and capable of determining for weal or woe the intellectual climate of physics. Were he alive in the middle of the twentieth century he would surely have something pertinent to say about his metaphysical reflexions on modern physics. Of course this is vain speculation, but it is surely illuminating that we find difficulty in trying to imagine Hertz in this context. Hertz expressed himself with such clarity about the basic difficulties of mechanics, and showed so much practical sense in his appreciation of the power of its highly evolved symbolism, that he could confidently be expected to seize with enthusiasm on the magnificent formal inventions of modern theory. But as an imaginative physicist he would hardly be expected to surrender his view of the actual world to the mystique of the formalists. This is the real point at issue in discussing philosophically what is written for the purpose of affecting man's view of physical nature.

So long as it is possible to present clear classical pictures of what has been observed and of the techniques of experimental discovery, there is valid communication to a wide public. The explorer returns with an account of his discoveries to share the new thing with his fellows. Nowhere is this better exemplified than in the impressive recent story of geophysical research. The exploration of space near the earth has shown how ignorant we were, and are of our environment. But in reporting this new knowledge we need not shock the reader or hearer with abstractions that appear to challenge the common bias of experience and language we all inherit. Strange things, new things, and the unexpected have been found in these explorations, just as in former centuries travellers like David Livingstone in Africa found and reported to their sovereign strange men and customs and fresh wonders of nature in far-off lands. In this same context some men wish to put the inventions of theorists.* A new way, they assert, altogether different from that to which we were accustomed, has been discovered. We are pressed to accept the new thing as a fresh marvel without pausing to ask how much our credulity is imposed on. This is the human situation which the philosopher is perfectly entitled to examine critically. Credulity is surely his business. And surely also, if at the end of his labours he has removed the aura of the supernatural from the report and helped men to see in a matter of fact fashion that what seemed strange and unacceptable can be understood in a reasonable way, his work has a practical value to men.

This is no easy task. He finds few listeners among physicists. Having been trained in the attitudes of mind and forms of expression through which the elaborate inventions of modern theoretical physics are conventionally approached, they have no taste for

* For example, B. Hoffman, *The Strange Story of the Quantum* (New York: Dover Publications Inc., 2nd ed., 1959).

philosophical questions that they have learned to avoid, because such questions appear to lead to no immediate practical issue in physics itself. Rather, philosophy is tolerated merely as a vehicle by which new physical theories are presented to astound the reader or hearer not professionally engaged with the science. In these revelations the fame of the scientist who dispenses intellectual shock usurps the authority that should derive from clear explanations in terms that are well understood, and our common tongue that should serve to communicate and join men in intellectual sympathy serves only to divide them in a babel of private meanings.

It is significant that Hertz had a better grasp of the role of informality in thinking about physics than is evident today among physicists. His allusion to the thoughtful mind having needs which we are accustomed to call metaphysical seems to indicate his awareness of the effect of social forces in the world of science. Yesterday's unorthodoxy becomes the accepted convention of today's practice, but physicists, and mathematicians, generally seem to have little understanding of how this is done. It is not the subject matter of physics: and since there is room here for variety of opinion and there are no easy formalisms to encompass the wide scope of human behaviour, it has little interest for them. Thus one arrives at the practical arrangement by which only duly formalised contributions to science are acceptable, without looking too closely at the grounds on which they are judged to be so. Clearly a really new idea must in some places challenge what is accepted. Somewhere it does not fit in. It must be nursed to give it a chance for effective life. If the inventor can point out succinctly where one can discriminate between his hypothesis and those he would replace, either he or another experimental physicist must commit himself to the labour of carrying out the tests.

This way of talking about the matter suggests that we have in mind a clearly delineated idea that is simply and logically presented in a special field of study. It does not apply to the processes by which we approach the 'deep questions'—as theoretical physicists and some mathematicians are wont to say in referring to difficulties in understanding—that touch our concept of nature.

Mathematical inventions that achieve a wide measure of success in their physical application have been evolved with a great deal of labour and ingenuity. Accordingly, most men are little disposed to accept light-heartedly philosophical criticisms of the orthodox physical interpretation when they are not elaborated constructively into predictions that can be tested by experiment. So far as the internal structure of physics is concerned, this attitude is a practical one and it justifies, in some degree at least, concern to defend physical theory from metaphysical attack.* Of course the essential defences of currently accepted physical theory are found at the level of the unformalised non-mathematical metaphysical ways of speaking and writing. These ways have been chosen by Bohr† and other leading physicists and their disciples, and their effect was originally to guard the early mathematical inventions which were known to work. The process going on here is much different from that presented by writers who wish to substitute for mere plausible reasoning the compulsions of academic formalised logic. By means of the latter, metaphysical comment is judged not on its own merit in contributing to understanding but set against the mathematical theory. In so far as metaphysical theses are advanced to question accepted physical interpretation of the theory, the metaphysical comment is unlikely to be received patiently and attentively. Besides, new ideas are elaborated mathematically in terms of accepted formalism so rapidly that the essential simplicity of metaphysical comment is overwhelmed. In the light of this it is remarkable that there should persist today substantial traces of dissatisfaction with the metaphysical doctrines of quantum theory. For they are metaphysical doctrines, inhering in the manner of speaking about the reality we investigate by the methods of physics.

Consider the phenomenological descriptions of the family relations between the numerous elementary particles we now know. A large body of physical information has been accumulated about

* See N. R. Hanson, *Amer. J. Phys.* **27**, no. 1, 1–15 (1959).
† R. Kronig, in *Theoretical Physics in the Twentieth Century*, pp. 31–2 (New York: Interscience Publishers Inc., 1960).

them in many ingenious experiments. To reduce this to order a sophisticated accounting system has been set up and theorists are engaged in trying to put something physically meaningful behind it. In principle this activity is not different from that which critics of some metaphysical forms of expression current in theoretical writing are condemned for engaging in. Are not these critics also concerned about physical significance?

In a lecture delivered in October 1960 at the University of Tokio, Professor L. Rosenfeld of Copenhagen discussed the foundations of quantum theory and complementarity.* He was concerned that 'Young physicists are raising doubts about the correctness of the basic ideas of quantum mechanics, and try to do it better', and added, 'These efforts are, I am afraid, rather futile, because they rest on a complete misunderstanding of the really very difficult conceptual situation which quantum mechanics presents to us'.

It is quite remarkable that Rosenfeld finds himself in a metaphysical discussion to support his view that in formulating 'complementarity' Bohr created a new logical implement and thus made it possible 'to surmount the logical contradiction that must arise if one tries to give universal validity to the concepts that are mutually exclusive'. If there is to be philosophical discussion of this matter, we might ask why we should be concerned with a logical principle to sanction what we do by second nature in our daily life, giving up one form of representation as we take up another. We are perfectly familiar with the limited applicability of the pictures we make.†

The reason that young men may be raising doubts about the correctness of the basic ideas of quantum mechanics is probably the simple one that they are dissatisfied with these ideas, at least as presented in accordance with current fashion. Quantum mechanics, during the nearly forty years of its existence, has achieved such great success and contributed so impressively to the

* *Nature, Lond.*, **190**, 384 (29 April 1961).
† See W. H. Watson, *On Understanding Physics*, ch. III (Cambridge University Press, 1938).

5

evolution of physics that one does not expect to rewrite its mathematics in any major sense. What is at issue is how we explain and justify on physical grounds the forms of representation that are treated by quantum mechanics, and how we should speak of the world that is revealed to us by experiments interpreted by its methods.

If more interest is being shown today than formerly in the foundations of quantum mechanics we need not be alarmed! From the invention of the differential calculus to the writing of the logically satisfactory expositions of its structure required more than a century. We must expect change and evolution of our intellectual attitudes to physical theory, especially through exercise of the imagination. Rosenfeld takes a formalist's view of the matter, his mind closed to the possibility that what is needed is to show the learner something more illuminating and persuasive than is found generally in the physical explanations that precede the presentation of mathematical formalities. This is a matter that will be decided in the course of time by the inventions of another generation, and they, no more than we, can compel those who come after to accept what is passed on.

We have to deal not with vague generality of concepts but with the physical relevance of waves and particles to the representation of radiation that consists of atoms. This is the logical situation inside physics that must be explored. It is a matter of exposing the anatomies of different representations and putting the ideas in their proper places, disciplining ourselves to hesitate in continuing to accept the metaphysical bias of ordinary language and our habits in using it.

Inventions in theoretical physics are justified not in their intrinsic mathematical form but in the useful life they may achieve in men's minds. Likewise metaphysical activity is justified *post hoc* by the way it serves men. It need not serve all men, but it can serve some thoughtful minds. Among these in our time we find the distinguished Nobel Prizeman in Physics, Louis de Broglie, who seems to lapse periodically from orthodoxy and return to his strong intuition about the physical existence of the entities of which

6

modern physics speaks. He is clearly dissatisfied in thinking about metaphysical questions associated with quantum mechanics and he has tried to construct a new theory ('Non-Linear Wave Mechanics') intended to avoid the metaphysical conflict. He has failed and it is surprising that he should have expected to succeed. The resolution of problems set by his intuition in the face of modern formalism in theoretical physics does not lie in physics. Indeed, the *mise-en-scène* is a good copy of that in which are found the philosophical questions generated by philosophers about ordinary language. To extricate ourselves from their obsessive attendance on our thinking we must look beyond the words to the behaviour of men who ignore the question because it does not stand in the way of their effective activity.

In the spirit and, no doubt, under the spell of the great French realist Pascal,* de Broglie knows that experimental physics deals with no figment of the imagination but with the real world in which we live. Physical theory must come to terms with the actualities on which we depend when we investigate nature. Accordingly, de Broglie is not disposed to accept the wave-particle duality without imagining a physical mechanism that can transport an electron, for example, from its source to the place where it is detected. But in following this inclination he has departed from the tenets of orthodox quantum mechanical teaching in essential particulars, and he has not succeeded in achieving the satisfaction he sought in his concept of physical nature. His concern to achieve an acceptable physical idea of the existence of elementary particles in space has kept alive his interest in the wave-particle paradox. Yet in the minds of learners of physics this paradox is supposed to be resolved by the formal exposition found in works on elementary quantum mechanics. It is remarkable that the man whose insight originated the concept of matter-waves remains dissatisfied with these explanations which, if they were really effective, should have removed his metaphysical difficulty. Why haven't they?

* See Ernest Mortimer, *Blaise Pascal* (London: Methuen and Co., 1959), especially in relation to Descartes.

Are we to suppose that this man with a lifetime dedicated to physics is obtuse or merely obstinate in his resistance* to the blandishments of academic persuasion? Surely there is something valid in his intuition, however mistaken he may be as to its proper place in relation to physics. Showing what that proper place is may not be so far off as the extreme formalists seem to think, for there is evidence in recent works on theoretical physics† of a trend to naïve realism but not, it is true, in the way de Broglie has tried. Although in the context of experimental physics his intuition about existence is well founded, it loses its effect in microphysical representation because the ordinary idea of existence in the world is indissolubly connected with classical modes of thinking.

Surely de Broglie, Bohm, and others are concerned about imagining something existing in the world. That 'no effective algebra'‡ has come from them in their attempts to satisfy their metaphysical yearning is really beside the point. Why are they dissatisfied? Orthodox quantum theorists have in mind that their theories are to apply to actual physical experiments; they are only too ready to claim that they are the true realists in that theory deals only with accounting anyway, and they do not pretend to calculate anything 'that cannot be measured' ('observed'). But this claim is in some respects misleading. How does the theorist know what can be measured? Is he prejudging the issue by this expression, or should he not say that the classical possibility for measurement is not presented in the forms of his representation? In so far as experience does not exhibit that the classical measurement is possible, his representation is more appropriate than the classical. If one wishes to give his method authoritative effect, this is done by the commitment to use it. Once committed we are no longer free. But the compulsion or necessity is on us, not on nature, to follow the method.

The metaphysical difficulty that has occupied the attention of de Broglie is just a part of the complex of metaphysical questions

* Cf. P. W. Bridgman, *Scientific American*, p. 201 (October 1960).

† Cf. W. Thirring, *Introduction to Quantum Electrodynamics* (New York and London: Academic Press, 1958).

‡ N. R. Hanson, *Amer. J. Phys.* **27**, no. 1, 1–15 (1959).

that arise in relation to the modern theory of quantum fields. There we speak not only of the creation and annihilation of particles that we can detect physically, but also of virtual particles whose existence is so transitory and localised that their activity can be inferred today only by subtle reasoning. Nevertheless, as experimental techniques evolve the scope of investigation may be extended to bring these matters to experimental examination in a more direct way; for this is how physics grows. In the course of this process physicists will continue to employ instruments and machines which they are sure exist in order to probe microphysical existences in contexts that classical ideas serve to represent only with great crudity, if at all.

Instead of placing in opposition our ideas of particles and waves to display the astounding paradox in which classical physics involves us, perhaps we could with advantage direct attention to the confusion that attends our use of classical forms of expression in discussing modern field theory. For the latter has in effect destroyed uncritical use of classical representation in microphysics. A metaphysical accommodation has to be reached in the relation of theoretical formality to the unformalised activities of the physicists who use it. It can be achieved only through better expositions in our textbooks of the relation of theoretical physics to work in the laboratory. In turn this will not happen until physicists recognise that formalism while a good servant is a poor master.

The contrast between the formalised and the unformalised can be well illustrated by what has to be done to use an automatic device intended to simulate the activity of the experimental scientist in measuring the record made by an instrument such as a recording spectrophotometer. Suppose the latter is used to obtain the intensity of the light emitted by a source, or absorbed in transmission through matter, from the blackening of the photographic plate on which the spectrum is photographed. The record of the varying electric current from the sensitive photo-electric device of the photometer with associated amplifying circuit shows the fluctuations due to noise which have to be averaged and sub-

9

tracted from the current recorded when light falls on the detector. Usually interest lies in the wavelengths corresponding to maximum intensity in emission, or minimum intensity in absorption. Where these peaks are clearly resolved, and when there is no theoretical ground for interpreting them as resulting from the superposition of the effects of several close peaks, the interpretation of the intensity *vs.* wavelength curve is fairly straightforward. If, however, the peaks are sufficiently close that the effects due to them extend over the same region of the spectrum, the experimenter has to exercise judgment in interpreting the record. What criteria is he to use to decide in arriving at the numbers that constitute his experimental results for comparison with theory, and become the basis for physical interpretation of the structure, for example of the emitting or absorbing molecules?

Using a particular theoretical model, he may employ a computing machine to calculate the spectrum he should observe with his apparatus, and he can arrange to have this spectrum plotted by means of the machine so as to compare it with the record drawn by the pen of the recorder. He may use the computer to produce a family of spectra and choose what he judges the best fit. Usually he has not committed himself in advance to particular rules that he undertakes to apply in making the choice. On the other hand, the whole of this process could be committed to the computer if he did specify completely in the programme for the computer the detailed criteria of a consistent system to which he is willing to adhere. Some experimenters are ready to carry through this formalisation of their activities, others are not ready to resign from intervening somewhere in the process. The justification of the attitude of the former is that his procedure is economical and makes possible carrying through consistently much larger investigations than possible for the latter. The justification of the man who reserves for the computing machine only numerical calculations, and for himself the examination of the experimental record, is that he will have greater confidence in the result. These are clearly matters of choice. Nevertheless, one can predict with some assurance that as general confidence grows in the automatic system, experimental scientists will come to rely on this arrangement when the circumstances of its use are favourable for their work. The completely automatic system couples the experimental measuring apparatus by electronic means to an output display through a computer that carries through all the formal steps for connecting the phenomenon with theoretical expectation.

Since physics depends on conventions and formalised methods in precise measurement, it is difficult to see how in the long run the unformalised judgment of the experimenter can be accepted. A discovery in science may depend, indeed often does depend, on unformalised judgment—'a flash of insight', 'a hunch', 'a happy guess'—but the evidence that will establish the confidence of other experimenters in the report of the discovery must conform in many important particulars with formalised criteria for measurements in experimental physics. These comments are not intended to justify the use of elaborate automatic devices in experimental physics without qualification. Just as it is necessary to arrange for the checking of arithmetical and logical calculations in a computer, it is also necessary to introduce checking procedures of a different kind to determine that the output of the computer makes physical sense. Whether or not such checks are made formal and programmed, or remain unformalised and depend entirely on our intelligent critical attitude in using the result is a practical matter. The choice depends on economy of effort. What we can formalise relatively easily we do. In any case we should always in the last resort depend on our continuing experience to warn us when our former reliance on the machine is misplaced.

The formalised involves our judgment in a different way from the unformalised activity, for we can only accept it or reject it, whereas the possibilities of accepting or rejecting unformalised details are individual and do not compromise our commitment to a formal system.

The major role of philosophical activity in relation to physics is to exhibit as clearly as possible what is taken for granted, but in doing so to avoid formalising the process. The basis of understanding is examining what we ourselves do. When we start postulating strange things about the world, justifying our choice not by explaining the matter properly but resorting to a mystique, we should beware. We know that our knowledge and experience are limited, but that is not reason enough for our being imposed on to accept formulas for interpreting successful technical inventions when the interpretation strains language and seems not to make sense to us.

In philosophy we are concerned sometimes in getting our bearings; we should ask ourselves 'bearings with respect to what?'. A philosophical attitude grows and evolves and elaborates itself through the experience of the philosopher. It is not a static body of doctrine. It is effective in some man's life otherwise it is a mere skeleton in the academic museum. The particular attitude is exemplified by philosophising, by exhibiting how the philosopher is disposed to proceed in asking questions and in attempting to disengage himself from them if the process of seeking an answer does not lead to fresh insight and understanding. The notion of 'systems of thought' appeals to the accountant in each of us because we can busy ourselves with nice schemes of organisation without involving ourselves in the experience of the insistent and sometimes awkward facts of life that stimulated the particular expression of thought we insist on attempting to classify. The accountant has in mind the formality of the auditor, the lawyer the formality of the court. In intellectual life these analogies mislead us so that we try to establish authority in various ways to make effective in the lives of other men the writing and formalism we favour. But it is quite obvious that most of these attempts to compel men do not work except in the classroom under the authority of the professor; and this has its proper use.

Accepting a new idea or a new method in physics is not a formalised process: there is no regular way to bring it about. For that reason we should expect to observe variety in the behaviour of physicists who are philosophically confused by forms of expression that tend to mislead us, such as 'what can be measured'. It has two senses. On the one hand it refers to physical possibility determined by the state of technology and the competence, imagination, ingenuity and skill of those who use it, and on the other hand it refers to logical possibility in one formal system as spoken of through a second which bears a certain formal analogy to the first. The elucidation of such matters derives its importance from the way we go on reacting to the expression. If we interpret 'what can be measured' in the first sense, clearly we are thereby prevented from understanding its effect in the second sense, and also likely to hesitate in committing ourselves to a new form of representing nature depending on a system of possibilities different from that to which we are accustomed.

Let us consider a different example of our dealing with language in physics. In quantum mechanics, electrons or neutrons, for example, are imagined to be emitted from sources, to undergo diffraction, and to be detected. There we take for granted the causal connexion between emission from the source and the event in which detection occurs, and this corresponds to the attitude of scientists and engineers in practice. To establish that events in which particles are detected are really due to radiation from the source in the way intended, the experimenter may remove the source or interpose shutters or absorbers, or improvise other alterations of the physical arrangement to find out where the radiation he cannot see is going. Of course the same considerations apply when the radiation from the source is transformed into radiations of different kinds when it bombards a piece of matter arranged as a target. However, in the modern theory of the processes by which such transformations are brought about, the individuality of the atomic entities is not necessarily preserved. We cannot apply labels in the classical way and we do not depend on the continuing existence of entities that enter the process and emerge from it. In practice we can usually circumscribe the region of space relevant to the phenomena in which these transformations are observed, but in the theory this physical view seems to be lost when the particles are symbolised by plane waves.

These thoughts should lead the philosopher concerned with them to recognise that the use of language in relation to atomic processes is complicated. Consider the expressions 'objects can be made

to interfere',* and 'one can talk of an amplitude that charac-
terises the object'.† Or again, ask the question 'under what circumstances would a physicist be ready to interpret the connexion
between two (or more) events as due to the passage of a particular
particle?'. This is not a mere rhetorical question. It is relevant to
the procedures and attitudes of mind by which bubble chamber
and emulsion photographs are examined and interpreted in high-
energy physics. The way in which things are done stands in striking
contrast to the formalisations about physics—operational definitions and the like, strangely reminiscent of the spurious regularisa-
tions of meaning in ordinary language at the hands of some
philosophers. We just have to accept that if a good experimental
physicist is presented with experimental records never before
imagined or seen, he will attempt to cope with them, and the process does not have the kind of legal force that writers about physics
not concerned to imagine the particular instance are so ready to improvise and impose by arbitrarily simple accounts of what goes on.

Much of the trouble in thinking about these things is that we do
not pay enough attention to the entry of ideas into actual physical
situations. This has been characteristic also of the academic approach to classical physics even when there is no need to introduce
atomic entities, when the phenomena are on a molar scale. Think
of observing the pattern made by the ice crystals on a puddle
frozen on the pavement. How was that pattern developed? A very
great deal of imagination is required to approach this question, and
by experiments to make some progress in understanding to what
extent the explanation one is inclined to offer is plausible. This is
the point; plausible reasoning plays a leading role in physics—as it
does in other sciences. When we attempt to put everything on a
formal deductive basis we should beware; sooner or later it will
involve us in formalising by over-simplification those situations in
which men have to use theory in conjunction with their training as
experimenters to cope with physical facts.

* R. P. Feynman, *Theory of Fundamental Processes*, p. 12 (New York: W. A.
Benjamin, 1961).
  † *Op. cit.* p. 18.

Philosophical questions have no intrinsic importance. Some questions are important for particular men because of the way in which the questions perplex them and deflect or obstruct them in going on with some other activity to which they are purposefully committed in life. For some men philosophical activity is a sort of intellectual adaptation to experience, learning the new and un-learning old attitudes of mind. This is the process that is referred to as understanding. When we say that we understand something we may be confident that we can manage any question offered to test our understanding, but our expectation may be disappointed. A particular question may lure us to pick it up in the wrong way so that we are unable to recognise its logical anatomy. Instead of making the spontaneous response, we hesitate to answer; we have to examine the question because we have no ready techniques for dealing with it. The answer we gave ten years ago may not be the answer we are ready to give today. Our philosophical activity is embedded in and surrounded by the context of life. It therefore serves needs in relation to a particular time. What can be said today that appears significant to affect men's minds when they are engaged in modern science may have little relevance for the next generation. Explanations that satisfy us sooner or later are super-seded. For explanation is a process that works relative to training, experience and 'the state of the art'. We cannot regularise this word any more than we can regularise the process of growing up because explaining is part of the business of living, of learning, and understanding.

CHAPTER II

IMAGINING WHAT IS GOING ON

Thirty years ago the neutron was discovered. At that time the alpha particles of radium or polonium served as the most useful probes to investigate the nuclei of atoms, because the million-volt accelerators that were to lay the foundations of nuclear technology had still to be developed. In several laboratories a penetrating radiation was observed to issue from the target when certain light elements were bombarded with alpha particles. It was thought to be gamma radiation of very high energy, so high indeed, that Chadwick regarded this interpretation as unreasonable, and proceeded to look for the recoil of nuclei hit by the radiation. He imagined it to consist of neutral particles knocked out of target nuclei by the alpha particles. He not only found recoiling protons and nitrogen nuclei, but by measuring the energies of their recoil, he was enabled to calculate the mass of the neutron he had discovered. The conclusive step in the demonstration was provided by cloud chamber photographs that showed tracks of recoiling protons. When Rutherford was shown Dee's photographs showing the recoiling nuclei, he is reported to have remarked to Chadwick, 'Do you mean to say that all this has been going on, and you didn't know it?'.

How often do experimental discoveries in science impress us with the large areas of *terra incognita* in our pictures of nature! We imagine nothing going on, because we have no clue to suspect it. Our representations have a basic physical innocence, until imagination coupled with technical ingenuity discloses how dull we were.

The study of cosmic rays was started by the observation that carefully constructed electroscopes are discharged even when enclosed in a mass of dense matter sufficiently thick to shield them from the ionising radiations that had been discovered at the turn of the century. Taking ionisation chambers up in balloons, lowering them into deep lakes, recording the ionisation current when the instrument is ship-borne round the earth all contributed to forming an early idea of the radiation that today is imagined by some scientists to originate in galactic explosions.

The past forty years has been a period of unprecedented human audacity in conceiving the universe in which we live. Even our concept of the Earth itself has been enlivened by theories that propose the creation of ocean basins by meteoric impact. The space between the Sun and the planets that once was an uninteresting void is now the arena in which radiation and matter from the Sun and mutually entrapping plasmas and magnetic fields are propagated to affect the life of man.

Imagining what is going on is the lure of young minds towards physics. It is also the concern of mature men who are moved by wonder and a modern appreciation of the drama of the Book of Job, in which man is brought face to face with a universe he did not make, awesome in its immensity and power, elaborated and articulated in detail, past accounting.

But what does it mean to imagine what is going on? Think of our imagining the nature of the damage to one of our limbs in an accident. Led by the onset of pain as we move it and touch it, we gradually localise and associate the pain with a particular muscle or tendon or bone. We explore a neighbourhood by movement guided by past experience of how the painful sensation depends on attitudes and movements involving the injured member. We even test our idea with our memory of the experience of falling to see if the injury we imagine is consistent with the physical cause. Imagining in this context is part of the process of adapting ourselves to a new (internal) environment. It is usually lively in response to real change in our condition. Such imagining is the basis of all practical initiative of men in their work. It goes on, and engages us with the future and what we shall do next. Most reasonable men regard formalising this behaviour as impossible in general. Nevertheless, techniques in diagnosis are known and used by physicians who have to learn what to ignore as irrelevant and what to look for. They proceed by testing ideas. This feedback process is the basis of the iteration towards confident diagnosis and effective action.

Look at Niagara Falls and then at a later time try to imagine it. The vividness of impression is confined to a few elements of form sufficiently persistent for us to discern them and remember them. We cannot usually achieve the effect of a good moving picture in colour to evoke the experience of our presence beside them. Here, imagining what is going on means to imagine the observed phenomenon. And this is always part of the scientific or engineering imagining. But the hydro-electric engineer in his work is not concerned to recapture his experience in this way alone. He is concerned with the amount of water flowing, with the height of the Falls and with the peculiarities of the flow that could affect the efficiency and endurance of the works he would instal to generate electric power from the Falls. For this purpose his imagining is a simple one. He does not consider the details of the cinematographic representation except in so far as they appear physically relevant to the engineering work. He is more likely to be interested in imagining the detailed mechanical and electrical functioning of the generators and the hydraulic functioning of spillways and turbines. This imagining is supported by the elaborate techniques of representation that have been evolved through the experience of many men and which the engineer had to learn in training for his profession. The work of many men is needed also to make the calculations, plans and models in preparation for the work of construction, for the amount of detail that must be encompassed is vast, and experience has taught that attention to detail is essential to engineering success. And it is so in physics also; but it must be detail relevant to our theoretical competence.

The child of mechanical bent, who dismantles a simple mechanism or electrical circuit and reassembles it correctly, has laid the foundation in this experience for imagining it functioning correctly, and when the mechanism fails to operate he may diagnose what has gone wrong.

Without any prior theoretical indoctrination each of us has

learned from experience to get a better view of things close up and to expect that by seeing detail with greater clarity of vision, we can understand them better, recognising how they will appear to us as we act on them. But this common expectation is not always fulfilled. Seeing in greater detail may merely confuse. We must have formed the right ideas to order our impressions in a coherent, significant way. And forming the right ideas is not trivial, as the history of man's learning attests. Men did not readily accept the round earth, the Copernican view of the Solar system, and Newton's laws, for example. We apprehend these ideas in learning our language and using it, in the early instruction in school and in the more advanced discipline of academic study of science through books and in the laboratory. We have to be shown how to take advantage of what has been learned by men in the past. Whereas in former times the scope of this instruction was for the most part confined to the office of the clerk—for in those days men of intellectual power interested in the pursuits of primitive science and technology learned under auspices separated from the ordinary affairs of educated men, today the young aspirant in science submits to a long training in mathematics and science before he engages in research in physics. During this process he acquires much more specialised and abstract ideas of what it means to imagine what is going on. He has learned certain forms of representation current among physicists. He speaks of these with the same ellipses and allusiveness of speech accepted among men in speaking of ordinary affairs. Through his training he knows how to improvise a particular invention for the special case and he takes for granted that his fellows can do the same. He learns a scientific vocabulary through use and, far more than is disclosed in the textbooks provided to instruct him, or those whom he himself will instruct, he depends on the experience he acquires informally as a learner, adapting himself intuitively to the life among students of science. He learns how to go on in physics by experience.

From the simple representations treated by elementary mathematics he proceeds to more elaborate theoretical invention consolidating his grasp and comprehension as he integrates what he has learned by association and by formal organisation based on mathematical abstractions. Through this experience the simple imagining of the working of a physical process is transformed. The simple models on which he once depended will serve only simple ends. He must be ready to look at the actual physical systems he wishes to represent in order to introduce the operation of physical causes ignored in the simple systems.

The simple model represents an actual physical machine or process only in certain respects. To elaborate the model into a more complex one in the expectation that a better representation will result, we examine the system represented to see what we have ignored and we imagine in the light of our experience of physics what processes we ought to consider relevant. In turn these may be represented only in a crude way; how crude depends on what the purpose of the enlarged representation happens to be. The scope of possible representation is determined by our ability to imagine something going on and therefore depends on the ideas we can bring to our aid to give form to our thoughts.

The classical resources of the modern physicist are the atomic theory of matter and the theories of processes of energy propagation and transformation in it. By training, the mathematical analysis of quantum mechanics supports the physicist's thinking in passing beyond the bounds of successful classical representation and has provided him with concepts that have to be explained mathematically. To be properly understood these concepts demand an adequate measure of mathematical training. Nevertheless, in experimental physics today just as in former times and in other activities of men, the teacher may present only an elementary account of these mathematical ideas intended to introduce them to the learner, and depend really on the experience of using them in a physical context to establish their significance.

An important part in the training of physicists is played by the elementary descriptions of basic experiments which are presented so as to acquaint the student with some notion of modern ideas by introducing them in historical order. In this way a vocabulary is learned under conditions resembling our learning the meaning and use of words and expressions in ordinary language. The discovery of the electron and the revelation of its properties in a variety of experiments establish in the mind of the learner some understanding of the family of circumstances related to the use of the word 'electron', so that when later he begins formal instruction in mathematical theory, this understanding is taken for granted. It is formally not represented in theory but it is part of the training of physicists enabling them to understand and apply theory in interpreting physical experiments.

In this way we pass easily from one level of theory to another, usually a more abstract one, gradually getting out of range of the physical basis of our thinking and ignoring our dependence on our bodies and the apparatus not only of the laboratory, but of life as a whole in doing physics. Thus mathematical abstraction can be offered without its having to face the question 'How is this relevant to the real world?'.

It is against this complex and sometimes confused background of unformalised behaviour that 'imagining what is going on' has to be given meaning. Clearly it means different things to different men, and it depends on what they have in mind to use it for. Let us not be tempted on that account to assert that it is 'meaningless' with the prejudice that multiplicity of meanings is synonymous with no meaning.

Consider the scattering of monochromatic X-rays by a crystal. From the measured distribution of the intensity of scattered radiation we proceed to infer crystalline structure and particular arrangement of the atoms on a lattice in space. We relate the intensity of reflexion to the imperfection of the lattice which we imagine composed of microcrystals and affected by the heat motion of the atoms. By other measurements and observations we can learn more about the crystal and with each fresh observational approach we may need to improve our representation. Side by side with the evolution of techniques for measurement and recording, there marches an evolution of techniques for representing the material structure being studied and the motions of its parts. These 'pictures' resemble maps some of which are readily interpreted and others, using uncommon projections, have to be explained and understood with the aid of mathematical discipline before they can be used. Roughly speaking, special representations have special applications. The picture which we are all ready to accept is that which seems to place the atoms of the crystal in the same physical space as ourselves, and moving in that space very much as we ourselves move in that space. In actual fact, however, while we can make good measurements of the average distance between the planes of atoms we are unable in practice to set the crystal in relation to its environment with the same degree of precision. We do not regard this as a limitation in principle so the classical view of the matter serves and avoids the complication of treating the crystal as located in an abstract space separate from that we can survey in the ordinary way. But of course there is nothing very remarkable about this circumstance, nor is its incidence peculiar to physics. Cartographers preparing maps for the motorist sometimes show out-of-scale enlargements of metropolitan areas on their map in order to increase its usefulness. It is quite recognised that a map showing detail can be applied effectively only in the context where what is shown in the map can be recognised on the ground, and on this basis a map intended to show a large area on the ground can be applied only crudely.

In imagining what is going on we are quite ready to pass from one representation in which we cannot show detail to another in which we can. Our imagining is limited by the scale of significant representation. Now in classical physical representation we seem free to make more and more refined pictures as we exhaust the possibility of each in turn. Consequently we are tempted to imagine structure and processes in the classical way with infinite refinement in principle. Against this we have to set the teaching of experience that matter consists of atoms, electric charge is quantised in integral multiples of the electron's charge and so on. Accordingly we are obliged sooner or later to interpret the sharp classical pictures in a different way. There is a physical limit to refinement in representing matter by the classical method. This limit is not sharp but a no-man's-land between atomic representation and the material continuum. In it classical sharpness and certainty dissolve into mists of statistical fluctuations that are revealed by the scattering of light and the Brownian motion. We are forced to develop new methods to examine matter on this scale of fineness, and with the aid of these inventions atomic and nuclear physics evolved. Their own proper methods of representation grew out of experience of quantum phenomena. Through the mathematical inventions of quantum mechanics a new way of representing what is going on established itself.

Learning how to use the methods of quantum mechanics in theoretical physics exhibits the same dependence on past experience that is evident in the laboratory. The student becomes familiar with certain important aspects of the methods employed in a process of gradual complication of the physical models treated by the formalism. He learns progressively of the mathematical inventions that have given form to new physical ideas as they emerged in the complex private and public activities of physicists engaged in research. The subject matter that is formalised is continually growing, the intellectual process of accretion resembling the solidification of a liquid. This is a suggestive metaphor because it brings to mind the inverse process of melting, which corresponds to formalism weakening and the matter returning to the melting-pot of unformalised activity. Perhaps a better analogy is provided by biological activity in the tissue of our bodies. The vital processes that formalistic embalming destroys are the life of all science. In ignoring their relevance for understanding physics we miss important insights. We have to recognise that imagining what is going on refers to a large family of related activities, some closely and others very distantly related to each other. No good intellectual purpose is served by trying to fix and preserve by the techniques of the taxidermist, or of the librarian, what must live in men's minds and be used by them in the evolution of thinking about physical nature for the purpose of exploring it, understanding it, and enlarging man's competence to cope with it. The main justification of formalisation is its economy for two main purposes. First, it enables the instructor to present ideas with minimum dependence on experience of the historical process by which the inventions to be explained happened to evolve. Secondly, by establishing the logical rules of the formalism axiomatically in a self-consistent scheme, the mathematician is enabled to carry out with confidence long chains and other more complicated patterns of inference. But not necessarily free from error—although with the advent of the large, fast computing machine men have now achieved impressive success in exploiting the mechanism of formalisms.

Calculations and other formal processes connected with representation are mere aids in imagining what is going on. They belong to engineering thoughts: they do not give thoughts significance. Nor do thoughts possess any occult property that could be given that name. Significance is revealed in what we do with them.

Dirac in his treatise,* *Quantum Mechanics*, expressed his attitude to pictures of microphysical processes.

What constitutes a picture? An image of a machine or a model made up of the primitive objects that serve for elementary instruction in classical physics? A representation that can be drawn on the blackboard and used to explain what is going on as a lure for thought, enabling the viewer to keep in mind connexions in the system represented? Such use of diagrams is an essential element in physics today as it has been in the past and there seems no disposition on the part of physicists to give it up. Dirac was concerned to have his reader refrain from relying on classical pictures at the microphysical level, because they distract from the essential connexions treated by the mathematical methods of quantum mechanics. These connexions are quite different from those of our classical method established by a long and successful experience in celestial mechanics, with the machines men have invented, and with the properties of matter relevant to engineering before it was invaded by the inventions based on atomic and nuclear physics.

The essential point about classical representation is that it is consistent with the experience of non-scientists and shares in the usages of ordinary language. The physicist's instruments and tools

---

* P. A. M. Dirac, *The Principles of Quantum Mechanics*, 4th ed., p. 10 (Oxford, 1957).

extend its range to larger magnitudes and also to smaller ones than we can apprehend by our unaided senses. In this context these instruments are mere practical aids for enabling us to survey our environment with greater discrimination and higher resolution, so that we can represent phenomena in the greater detail for which the continuous spaces conventionally used for representation leave a place, indeed present the possibilities.

In the physics of the twentieth century the methods and apparatus used by experimenters in the laboratory continue to serve the same purpose as their forerunners did in the nineteenth for classical representation. They relate physical phenomena at the atomic level to the world of ordinary language. But the representation no longer presents the possibility of indefinite refinement by the classical method. It is limited by the existence of atoms and the forms of phenomena that quantum mechanics was invented to manage.

During the past half century as ideas of new ways of conceiving atomic mechanics were developed, new ways of investigating phenomena evolved. The results of experiments could be interpreted satisfactorily ('correctly' according to the view accepted today) only in terms of the methods of representation based on quantum mechanics. The fact that the experimenter is still using apparatus he can see and handle in a public, objective way, and that can be described so that other men can repeat his work, supplies the connexion between the new representations conformable with our experience of quantum phenomena, and the classical gross representations and images of our environment, and naturally also the ordinary language by which we speak of it.

We use 'ordinary matter', objects of sense, fabricated often in new ways by modern technology, disposed in arrangements that never would have been conceived if physics had depended only on classical ideas. These are employed with an ingenuity in invention today beside which the work of Rutherford and other great innovators during the first two decades of this century seems primitive. Elaborate techniques to isolate the objects of study, high-speed counting, and as a necessary concomitant, fantastic precision in timing and in other physical measuring operations are in wide use today. Perhaps most significant of all, data-processing and computing aids enlarge the scope of the information that can be effectively garnered and organised for study and interpretation, and of course, preparing further steps in experimental investigation. All of these technical innovations distinguish modern experimental physics from the activities that preceded it.

For the purpose of imagining what is going on the experimenter is not restricted to a monistic unified system of representation. This is a fact that mathematicians and philosophers tend to overlook in their anxiety to achieve a simple formalisation of what goes on in physical science. How the physicist thinks of his experiment depends on what he has in mind to do. In the preliminary adjustments, calibrations, and tests of the functioning of his equipment he is concerned usually with matters that can be quite well managed in gross classical terms.

It is only when he has to think about what is going on at the microphysical level for the purpose of understanding what is happening in his instruments, either to interpret measurements or to diagnose malfunction of the arrangement he has set up, that he is forced to introduce the concepts and methods he learned through his training in quantum mechanics.

This training showed him how to use a mathematical theory and how to conceive of experimental investigations of microphysics for the purpose of applying the theory. By this training he learned a new language for speaking, writing and thinking of the operations in the laboratory by which physical research is carried on. He learned how to use diagrams to explain and support his thoughts just as he had to learn to use diagrams in his earliest contacts with physics. The rules for drawing and using these diagrams differ from those applied in classical physics. But a particular classical use of a diagram may be consistent with these rules and therein lies a possibility for philosophical confusion. To the legal formalistic mind this flexibility in using diagrams is intolerable. To the active physicist it is an advantage: he can exploit its economy by the informality of his behaviour in relation to it. Academic rigidity is out of place in the laboratory.

The processes just alluded to are in principle no different from what has gone on in the past when men have acquired a new technique. The technique is invented, is shown to a few who are capable of appreciating it; they use it, help to develop it, and eventually the successful technique has to be shown to learners. The conventions of rationalisation may enter at this stage, but unless they contribute to economy in instructing they are unlikely to be popular. The effective learning is by imitation. Indeed, this is the only way to learn unformalised behaviour. That this learning does not conform to the prejudice of bookish men is unimportant in practice, although it may be intellectually significant for the student of men. How the learning process goes on is a matter for study by biological scientists. Our minds should now be sufficiently open to receive the possibility that this natural process works, to accept it as an essential element in transmitting knowledge and understanding how it is used, and also (this is philosophically important) as an essential element supporting the formalism of mathematical theory. Without the training, the connexions between the mathematical theory and experimental practice would be absent. The training is partly formalised so that it may on occasion become a mere performance by the instructor, but this picture of training which is seductive to the unwary, does not correspond with essential experience of it. The instructor is able to cope with questions and responsive behaviour from his pupils which he has no programmes for dealing with in advance. In any case how does he know which programme to apply even if he has prepared them? His behaviour responds to the behaviour of his class, if he is a good teacher. Without the idea of feedback from the learner we do not go far in understanding what is going on. These comments are not intended to explain imitative learning, they are directed merely to discourage formalistic objections so dear to academic philosophers in the form 'This cannot happen: it is impossible' and elaborated through verbal logic grounded in a tradition ignorant of the potentiality of biological systems to react adaptively to their environment.

Thus we reach the view that imagining what is going on depends not only on learning mathematical systems of representing nature, but also on the training by which physicists have learned to communicate with each other about their experience and to connect the concepts and mathematical training in using them with the apparatus, processes and behaviour of the physicist in the laboratory. The structure of the mathematical theory resembles the rules of chess. In order to play the game we have to learn the names of the pieces, to recognise them and so on, and are ready to improvise so as to continue the game even should an accident damage a piece, for instance.

In stating these things, we have merely recognised that physicists are able to imagine what is going on in microphysical phenomena in ways that most of them seem to find satisfactory for their work. In so far as physicists find current inventions and methods failing to serve their purposes, one may look with confidence for new inventions intended to do better. Following the example of David Hume, philosophers must accept these as the facts of life and adapt their thinking to accommodate them. If they are in difficulty here, so much the worse for philosophy.

It has already been mentioned that imagining what is going on is not simple. There is no standard. Even in quantum mechanical representation one finds this variety. The representation shows some regard to the purpose for which it is to be used. However, compared with these relatively minor variations, the distinction between what is called 'elementary quantum mechanics' and quantum field theory is so great* that imagining what is going on by these two methods seems to require special attention. As usual the mathematicians are determined to promote the most elaborate theory they know as the correct theory and in this case the evidence is strongly in their favour, since the field theory dealing with electrodynamics has been used successfully to compute refined structure in atomic spectra that the 'elementary theory' failed to represent correctly. The important physical difference between the two is concerned with how we think about atomic particles and their interactions. In the field theory the creation and annihilation

* See Dirac, *The Principles of Quantum Mechanics*, 4th ed., Preface.

of particles are essential elements dictated by making the theory conform to the theory of relativity.

The quantum mechanics of the 'elementary' textbooks has its own set of philosophical questions connected with imagining what is going on. They differ in important respects from the questions that are related to field theory.

In the latter we are not usually concerned to trace the particles of each field as if we were looking at the track of ionisation produced by a proton, for example in a cloud chamber, calling it the same proton that issued from the target we bombard to cause the nuclear reaction of which protons are one of the products. Nevertheless, in the theory, the issuing proton is represented to have at a very great distance from the target an experimentally well-defined direction of motion and to possess definite kinetic energy. The bombarded nucleus loses a proton and the proton field undergoes a transition in which the appropriate proton momentum state previously empty now has an occupant. When a neutron undergoes β-decay and is transformed into a proton, electron and antineutrino, all three of these particles may be regarded as created while the neutron is annihilated. However, when, as is done in modern theory, the neutron and proton are regarded as different dynamical states of a nucleon, the neutron and proton fields are particular states of the nucleon field which undergoes a transition in replacing a neutron by a proton and acquiring one unit of electric charge.

The quantum field theory stands at least one step farther away from the classical particle theory than does the non-relativistic quantum mechanics. In the latter we are encouraged to imagine that the metaphysical presumptions of ordinary language are all right provided that we do not attempt to specify the motion with the precision of classical mechanics and that instead we accept the doctrines of the 'Copenhagen school'. Electrons and other entities of modern physics exist all right, but the mode of their existence depends on the physical processes in which they participate and what properties we set up apparatus to measure. We think of them as things entering our detector and in their passage through classical fields, but in representing how their passage is affected by the presence of other things we have recourse to optical imagery and the well-elaborated apparatus of physical theory for treating waves.

However naïve our thoughts of these matters, we must, nevertheless, distinguish between physical conventions in representing nature and the reality itself.

Whether we observe it or not, something is going on. Calculating probability distributions and thermodynamic averages is one thing. Imagining the particular actuality is something else. Think of the cloud chamber and the photographic plate, and recall to mind the drastic effects that have arisen from radiation and matter emitted by the Sun. These are not statistical nonentities. And reality is not composed of nonentities.

It is time to reassert as basic scientific belief that our mathematical theories are connected with accounting, not with the substance of reality. Indifference to chemistry and complexity of material structure, which can all be imagined, is indicative of limited scientific interest.

When Kepler discovered his laws he was not satisfied with this achievement; he had intuitions of the possibility of a mechanical explanation of them. What mathematical invention does in science is to move the boundaries between formal and unformalised thinking. Let us not assume that formalism bounds effective scientific thinking. It does not.

CHAPTER III

OUR METAPHYSICAL DEPENDENCE
ON LANGUAGE

The opinion that metaphysics has no place in the formal development and literature of physical science is widely held by physicists. Nevertheless, metaphysical bias is evident in that literature especially in theoretical writing about physics. It obtrudes from the prefaces of treatises and whenever a new approach to physical representation is being presented. For it is fashionable to present theoretical ideas as if each of them carried some kind of guarantee of its own effectiveness *a priori*. A writer reveals his metaphysical need to justify the commitment he has made, or to assert his confidence in the theory he treats, and so on. To dismiss these evidences that physics is the work of men, saying that, of course, such comments are psychological and subjective,* and not really part of physics, which deals with the objective, is itself a reflexion of metaphysical attitude. To admit that physics is concerned with the objective, however, does not blunt the point of the comment just made, for physicists depend on the attitude in their work. One merely accepts that this is how physicists treat these matters on which there is variety of opinion. The objectivity of physics is attested by the measure of agreement among physicists about established experimental facts, about the reliability of methods used to observe nature, and about the forms of theory that support experimental technique, determine how we represent physical phenomena, and provide the bases both for our consolidation of knowledge and for attaching significance to physical discoveries.

As to metaphysics, however, we are confused by the intricate network of unformalised adjustments on which we depend by habit, and by our intuition or inclination to follow the paths we do

* As Bridgman, for example, said of de Broglie in his review of the latter's 'Non-linear Wave Mechanics', *Scientific American*, p. 206 (October 1960).

follow in using our mother tongue. Our language comes to us inextricably interwoven with our daily lives and with a great literature that speaks out of the past not merely to men's minds but also to their hearts. Its influence is as varied as the moods and experiences in the lives of the men whose genius created the art that can so deeply affect us. Our learning of language and appreciation of literature come to us without our being critically aware of what was going on while it happened; and we are likewise inattentive to the changes by which language and behaviour evolve during our lifetime.

The history of the meaning and use of particular words and expressions reveals many interesting examples of the manner in which social changes affect them. Science no less than custom has influenced our language in many subtle ways which, like the growth of weeds, defy the attention of the academic gardeners. These things are really beyond our formal control. However much the philosopher may desire to regularise the informalities of ordinary speech, he cannot compel men to conform to the special ways he has in mind. His formal inventions are in much the same condition as formal inventions in science and engineering and in the practical affairs of men. They achieve currency by practical effectiveness. Whatever serves the needs of men will eventually be used. Inventions concerning language have to face the same criterion. Since, however, modern philosophers rarely have in mind to contribute practical inventions that might be expected to improve the lot of their fellows in the ordinary walks of life, their inventions usually achieve only a temporary private existence in the world of teachers of philosophy. Of course there have been some notable exceptions. Today, professional philosophy still lives on the estate cultivated mainly with the primitive implements of the dominie and the monk. In spite of the intellectual challenge facing modern man that should liberate him completely from the restraints of medieval formalism, and that has liberated him in science, everyone of us is cribbed, cabined and confined by our training and habit through the unformalised use of ordinary language.

The theory of evolution and a century of magnificent growth of biological science, the theory of relativity and the atomic theory of matter and radiation have each challenged aspects of our common assumptions about the fixed things in men's lives. Not only so, in discovery after discovery in this century, myths cherished in some of the most fondly remembered prose and poetry that have come down to us from men who lived imaginatively and adventurously long ago have been destroyed. Man's conception of himself in relation to the universe has been transformed in the lifetime of men living today. Through communication to all who care to look or listen, the engineering inventions that embody physical knowledge gained in less than a century are bringing to the mass of men some appreciation of the broadening of man's idea of himself through scientific insight. It is surely not too much to expect that soon these influences will also have changed habits of speech whose imagery once was primitive science.

With the growth of techniques for engineering in its broadest sense, our language has been influenced in many deep and far-reaching ways to incorporate the grammar of new ideas and enlarge its vocabulary with new names. The process is continually operative, transforming the fabric of language for the race just as the tissue of our bodies is renewed. At every point the teaching of science induces one to look through new metaphors at man's work, his activity in learning, his inventions and so on.

On the other hand we seem to have inherited a metaphysical theory associated with our language. It inheres in the formality of words and is tied effectively only to the early activities of our era in the writing of books and in formal teaching, not twentieth-century scientific understanding. It might be called the metaphysics of the grammarian, or the lawyer. From the formalities of these two occupations have descended various academic formalisms. 62232

A legal document attempts to be complete. It supplies its own special definitions and presents itself as the legal instrument of some executive action. Some parties are attached by personal interest to its effect. So long as they accept the consequences of the action prescribed by the document, there is occasion neither to resort to litigation to determine its meaning in conformity with law, nor to enlist authority to enforce it. In determining meaning, men refer to dictionaries, to common usage in life, to precedents in court records, and so on. The process in court is directed to resolving how some men should act. But this is not necessarily a simple closed process. One cannot tell beforehand how it will terminate, for that depends on how men do act. Action in accordance with the decision of a court may close the matter, but it may be opened again by someone who thinks that what has been done contravenes the judgment. Thus the conclusive appearance of the document does not necessarily portray life.

What has the law to do with physics? It is not in the dialogue of the play, it is true, but it is certainly implied in the stage directions. Physical laws have been contrasted* with legislative instruments to indicate how the use of the word 'law' outside science infects its associations in physics. Not merely the word, but how we behave in the social coming and going, attending to our personal affairs, is entangled with it. We are ready to use grammatical forms appropriate to the operation of the law, and to guide our thinking and our behaviour in obedience to their logic even when we are free from legal business.

One can point to other common sources of metaphysical bias derived from the experience and evolution of society recorded by social historians.

Our inclination to follow a particular metaphysical bias is more often than not associated with epistemological formalism of some sort or another. We rely on some theory, even imperfectly formalised conceptions as to how we think, and in our desire to complete the job, establish its categories as exhaustive. Whereas what is needed is to examine what goes on in the particular instance of metaphysical decision, leaving to biological scientists to discover the mechanism by which observed behaviour evolves. Men use ideas developed in one context to aid them in others, thus changing

* W. H. Watson, *On Understanding Physics* (Cambridge University Press, 1938).

the scope of their formal competence. When a scientific discovery is announced or an engineering invention exhibited and properly explained, our concept of what is possible is usually changed in some respects. In using our language we acknowledge the new thing and take it into account sometimes in a great variety of ways, sometimes with comic effect.

Our language itself has evolved, serving in successive eras in differing ways. Inevitably primitive science has trailed along with it. Primitive ideas of the internal functions of our bodies have been modelled on the processes that men observed in nature and in the practice of their fellows. As civilisation evolved the same sources presumably served to build up progressively a more complicated vocabulary. By reflexion men have elaborated subtleties of expression with a virtuosity whose biological roots we only too dimly perceive today. Somewhere along the way men learned the value of abstract nouns, at first with occult implications but later without them, and in our time we are still philosophising about them very much as Augustine was so long ago. Mathematicians have exposed the anatomy of 'number', but physicists have made little progress with 'substance' and 'existence'. They are quite content to rely on the metaphysical intuitions of ordinary language to give these words life.

It is remarkable how physics has left far behind our scientifically unschooled tendency to reserve the word 'substance' for condensed matter, indeed sometimes to reserve it for solid matter. Certainly, gas, though a form of matter, seems not to fit well with the common usage, and perhaps the fact that a liquid has normally to be held in a container limits our readiness to admit it as a completely satisfactory embodiment of substantiality.

We wish to say 'substance exists' to express our conviction that it has some sort of permanence and independence of our particular view of it. It is not merely imagined. It can be exhibited to other men who can touch it, see it, and so on. In our ordinary language the substance of a body is invariant with respect to the vicissitudes of changed form. Nevertheless, we know perfectly well in modern physics that this view of things is a crude one. Matter is physically and chemically affected even by the mechanical processes used to give it a particular shape.

In the laboratory the physicist depends on the metaphysical attitudes taken for granted in the naïve use of ordinary language. Instruments and apparatus are brought from the workshop where they may have been constructed or unpacked and assembled. They are installed in accordance with the plans for the work to be done. Eventually their operation is tested. Through this process the experimenter is in no doubt whatever that once having tested the equipment, he is free to turn his attention to its use in the investigation he has in mind, without having to keep the equipment under constant surveillance. Unless in the course of his work his experience leads him to doubt that all is well, he proceeds on the same assumptions that are relied on in the service garage, in the large engineering project, and in the simple affairs of the kitchen. This elementary matter-of-fact behaviour is an illuminating instance of metaphysical confidence. We might be tempted to argue that since the experimenter does not know for how long his test guarantees the effectiveness of his equipment, he is operating on conjecture. He cannot be sure that all is well. This attitude to practical common sense is found in some philosophy and ignores plausible reasoning in science* as in the ordinary affairs of life. Even if we laid down a continuous test of efficacy intended to guarantee that the apparatus functions as it was intended to, we can still raise doubts about other connexions we depend on in thinking and acting. It seems that the point of the philosopher's question is merely to establish the possibility of doubt and thus eliminate certainty. What men of one generation are pleased to regard as satisfactorily settled may be questioned by a succeeding one, so the philosopher is not disposed to accept the confidence of the scientist with his enthusiasm. What he overlooks of course is the variety of experience and multiplicity of evidence on which this confidence is grounded, and how in the light of evolving experience men may disengage themselves from a commitment in life and pay

* See G. Polya, *Mathematics and Plausible Reasoning* (London: Oxford University Press, 1954).

36

close attention to something they have previously taken for granted.

These thoughts could lead us quite naturally to discuss correctness of calculation which was taken up at length in Wittgenstein's posthumous works.* Viewed in face of the practical achievements with modern computing machines, the philosopher's demand for the limit of absolute certainty has a remarkable affinity with the demand by the teacher for the correct answer to a textbook problem. Here the number of steps is so small that the checking process can be accomplished easily. But this is not a suitable picture of how men base their judgment in real life. They are ready to learn from experience as they go along. This is the kind of process that goes on in science.

But there is no unique form for these processes; in the very large family of them, there are resemblances between some of them but not common to all. It is remarkable how ready some men have been to attempt to formalise linguistic usage as if it could be treated according to the formula: 'things with a common name must have certain characteristics in common', which we often use in classifying and labelling articles of commerce, but which we do not rely on in naming men. The same surname carries no guarantee that its bearers, even in the same family, possess a unique set of human characteristics in common. There may be certain family resemblances between some members but not between all, and there may be resemblances in different respects between others. Our earliest experience usually shows us this; nevertheless, we seem to forget it and, following the simple approach of the grammarian or lexicographer, submit to the fiction of the formula.

The variety of life resists our simple schemes to formalise our language and behaviour. Yet physics emerged from just such schemes, depending on ordinary language to lead us into the formalities of representing mathematically the results of physical measurements.

The use of language is very complex. It depends not merely on

* L. Wittgenstein, *Philosophical Investigations* (Oxford: Blackwell, 1953); *Remarks on the Foundations of Mathematics* (Oxford: Blackwell, 1956).

the intricate mechanism of our central nervous system, but also on our capacity to learn, based on biological habit and on our experience of life. It might appear that since physics is not concerned with biology these considerations are irrelevant for understanding physics. And from one point of view this is correct. Our basic difficulty lies in disentangling ourselves from our bias that simple models of the use of words and the behaviour that goes with them must serve to portray 'these elementary matters'. The simplicity—and therefore economy—of common expressions deceives us because we ignore the experience in our youth by which we learned them, and because we do not address ourselves to attempt imagining scientifically the complex mechanisms and varieties of biophysical and biochemical processes in our bodies that could make that experience effective in our automatic response to language. The intrinsic simplicity of formalism misleads us to believe that the written or spoken word or the graphic sign is all that is necessary. If we understand it, and we have the training appropriate to the circumstances of its use, it *is* all that is necessary. Nevertheless, we are tempted to think that the word itself should present all the mechanism required. Of course it does not.

Much of our readiness to embrace simplicity in representation is connected with our inability to cope with a more complex picture. Today we can see on every hand the elaboration of automatic devices that give man a vaster scope than ever before in representing things and operating on his environment. These inventions are added to the apparatus man is endowed with and has grown up with. They enlarge his concept of the role of practical action in relation to formal accounting and the associated logical operations by which technical progress is achieved.

It is against this background that we have to exhibit the words 'exist' and 'existence', and to examine expressions such as 'to give the grounds for belief in the existence of...' which bring to mind that something more than logic is involved here. Applied to an object that we can name, handle and see, speaking of it as existing seems to say nothing. The expression does have significance, however, when we are concerned in exploring our environment that we may be deceived by an appearance, due perhaps to hallucination (observations by other men are relevant here) or due to an optical illusion, or to our being duped by unfamiliar sensations of touch and weight. That is, 'exist' is meaningful in a context where 'not exist' is also possible before the evidence provided by further investigation settles which is correct.

The history of maps and map-making is illuminating for our purpose here. Early maps look so strange to us that it is easy to recognise how much our idea of the surface of the Earth differs from that of our remote ancestors. The social fabric which maps helped to create by providing effective technical aid for man's planning and doing also differed from that today. All of this is relevant in a variety of ways to changing the habits of men and thereby altering the support for the expanding use of language. It contributes to metaphysical presumptions we depend on in physics. Sophistication and subtlety in the invention of symbols cannot free us from our dependence ultimately on the informality of ordinary behaviour in using them. This is a logical prohibition because the formal is finite and closed, restricted as it is by our resources in space, time and matter to represent everything that must be recorded if the formalism is to work. On the other hand, the informal is not worked out in detail beforehand. It is not something that can be set out like the plans of a building, the design of an industrial plant, or the programme for a computing machine, because we choose not to commit ourselves in advance. The informal has to be elaborated in a living process. In the course of this experience men use tools, symbols and techniques in innumerable improvisations to deal with the unexpected details of particular occasions. To philosophise as if this complexity could be ignored and a representation of the whole process be 'set down in a book' is the great academic illusion. It affects philosophising about physics, for in a book it is very much simpler to use mere names and let the processes by which meaning is brought to light look after themselves.

The invention of writing with its automated mass-copier the printing press has affected men's thinking also. After all, the immediate things in this context are the written or printed words themselves, not what they stand for. So it has come about that the categories of grammatical and literary analysis play important roles in philosophical discussion. This is an unfortunate development: it has become a custom that feeds the mechanical monster. From the closed loops of formal verbal connexion we do not escape to the sources of thought in our lives unless we imitate the poet, or dramatist, or novelist who stimulates living response in us. It is otherwise when we meet the exigencies of life and act in anticipation of our needs to cope with our circumstances, or in response to new and perhaps unexpected stimuli. Then we have to break out of the verbal circle. In philosophising about 'existence' we have to do likewise.

A feature whose location is reprentedes in a map might not exist. We have to look on the ground at the place marked in the map to settle this. In the early days of exploring the North American continent, this process actually happened; it was important for practical commitments. The tests by which existence or non-existence is established work here by the direct evidence of our senses. In physics, however, we explore the world inaccessible to them directly, using refined instruments and elaborate machines, and with the aid of our theories we interpret what we observe. Because the train of connexions between what is immediately presented to us and what we infer is often long and complex, our instinct is to regard this with less confidence than we do the ordinary objects we encounter in everyday life.

We are thus lured to the philosophical doubt about the reliability of the procedures of science. We look for guarantees in advance— as if our method should work as a machine and be incapable of malfunction. Now in fact we do approach scientific investigation with real confidence in the methods we depend on. They have been developed from the crude ones that served in the past for the early exploration of the phenomena we have to deal with, and they are now elaborated through the experience and imagination of many men into well-engineered techniques. Past experience provides part of the guarantee, but in new territory, or in the evolution of a new method, the guarantee is not yet to hand. We have to rely on current experience to guide us, and in due course achieve a measure of confidence in our interpretation of experiments in the terms we have imagined, sufficiently good to free us in practice from the uncertainty of the philosopher. That does not imply that we refuse to acknowledge the possibility of future experience that may lead us to change our view of things.

Our confidence in a complex process or machine that stands between us and nature resembles in many respects our confidence in clocks, aeroplanes and the apparatus of modern civilisation generally. We depend on experience and plausible reasoning and are ready to change our attitude, improvise special arrangements, and adapt ourselves to the unexpected. The process of establishing scientific knowledge is a continuing one, rather like maintaining a system of highways in networks of minor repairs, by-passes, new bridges and new main highways. One does not wish to regard a highway system as finished and determined for all time; it evolves in ways consistent with the geographical features of the area it serves, the important changes being topological ones. Maps have to be redrawn. The give-and-take with life illustrated in the history of such a utility is hardly shown at all in the map, which is a still picture. The techniques of cinematographic animation have provided better means for showing change. This successful application should remind us how much the structure of our formal thinking depends on the methods we can use to show structure. Much metaphysical bias is connected with primitive techniques for representation. Whatever the technique cannot manage presents a problem.

In physics we ought not to have to face any difficulty in this respect, for the history of the subject records many felicitous inventions for representing nature. Nevertheless, in so far as physicists depend on ordinary language in discussing how new ideas must be interwoven with their activity in physics, and to do this must step outside the formal treatment of the subject, we should expect to find variety of metaphysical bias there. But these matters find only an irregular place in the literature of physics.

New formalism is presented as a closed system based on a set of axioms and justified by its success in physical research. Distinctions respecting how we should regard and talk about what we are doing in physics have to win their way, sometimes in a slow evolution that culminates in formal innovations in textbooks and treatises, but generally they affect physical thinking through unformalised discussion and spread like a mild infection in the communications by which men learn from each other in friendly discourse.

The contrast between the regularity of the formal and the irregularity of the unformalised seems to dismay us. We seem to be able to get on with physics, but we really do not know how we do it—at least we are sure that scholarly formulations intended to explain the matter to us seem to lead to no good practical issue in the only way that would satisfy us, namely, to supply new ideas for more effective physical thinking.

If we look back to Hertz, Boltzmann, Russell, Whitehead and Eddington, each of whom thought a very great deal about philosophical questions connected with physics, how little of what they wrote has intellectual relevance and 'pith and sense' for the problems of physics today. And we can find modern writers who with an even greater dependence on formality set out ostensibly to regularise our basic ideas so as to compel us logically to follow modern theory because of the new metaphysical bias they present and their work seems to make only a scholarly impact.*

To understand what men do successfully in the complicated evolution of modern science, our need is not formalism to justify it logically. Clearly that does not help, for the formalism is already committed. What we need is illumination about the kind of processes we depend on more or less involuntarily in behaving as we do in the laboratory and in theorising about it. Once we grasp that there is a mechanism of thought which is not represented at all in our language, any more than the programme for a calculation by means of a computing machine represents the mechanism and

---

* For example, J. Schwinger, *Proc. Nat. Acad. Sci. Wash.*, **45**, no. 10, 1542–3 (1959).

electrical functioning of the computer, we shall be freed from our metaphysical dependence on forms of presenting ideas that appear to convert us into mere observers of the processes that cannot go on without our active participation.

We cannot successfully explain a formalism merely in its own terms. We have to approach it through the informality of ordinary language, and through the informalities of learning. The latter is relevant not merely as it takes place on the particular occasion when the formalism is being studied. Learning in the past comes into account also because it is the necessary preparation to enable one to read, or hear, or understand new ideas. For the mechanism that brings words and diagrams to life does not work by magic. It depends on living processes that transcend our present scientific ability to analyse them in detail. But we are not ignorant of some understanding of them, and are no longer rebuffed as our ancestors were by our inability to imagine how they might work.*

We have to extricate ourselves from inappropriate metaphysical bias by inventing illustrations and examples using the apparatus that is well known to us in ordinary life. We may impose on its functioning the logical form we desire in order to exemplify a new idea, and imagine the mechanism made inaccessible to our observation, so that we are placed in the position of the observer of natural phenomena. We ask ourselves questions. What hypotheses would we propose to explain its functioning in the way it does? Would we invent entities that are not present in fact? Clearly we might. In so doing we should reveal to ourselves that the necessity we want to impose in our theoretical approach to phenomena cannot be applied to the existence of our invention. The model serves very well to remove this temptation. It compels us to look on our use of the words 'exist' and 'existence' in a way different from that associated with naïve dependence on ordinary language, which does not warn us explicitly of the varieties of the use of these words, or of our own participation in giving them meaning when we apply them.

* Cf., for example, W. Ross Ashby, *Design for a Brain* (London: Chapman and Hall, 1960).

43

We should note that there is nothing wrong with the metaphysical bias of ordinary language. Its usefulness is supported by long experience. What is at issue is relying on it uncritically in physics which explores new fields of phenomena. There, we no longer have the assurance that the logical forms given us in language are appropriate to the new experience.

In ordinary life we are ready to allow certain properties of an object to be altered and still use the same name. Is this a good practice in physics? Or consider another example of metaphysical bias from language: we regard a thing as remaining the same as we look at it in different ways. This metaphysical presumption pervades the whole of science. The justification for our relying on it is not logical necessity, but the extremely wide freedom it gives the investigator in devising new methods of studying nature. Nevertheless, in biological research, it is well recognised that living matter, being extremely sensitive to change in its environment, may be significantly altered by the process used to study it. Even in physics one is concerned with this same limitation, for example investigating the optical absorption by a crystal of alkali halide in general alters the spectrum of the crystal significantly. The spectrum depends on the previous history of the crystal. But such experiences are not regarded as of great philosophical importance. Scientists alter their methods as they learn from experience, as they must in order to make progress in understanding.

It is when we are thwarted in overcoming a difficulty in reconciling the evidence from different means of investigation that it is desirable to examine the metaphysical assumptions we have taken for granted. If we find that giving up one we commonly rely on gets us out of our difficulty and allows scientific progress to continue, we must resist the temptation to say that the bias of ordinary language is wrong, as if in the process some transcendental discovery has been made. The metaphysical bias of ordinary language is consistent with the experience in which language developed. It is only grammarians, logicians, and philosophers who want to impose formalised versions of this bias as necessary forms of our thinking irrespective of what our thinking applies to. It should not

44

surprise us that modern experience through the advance of science brings men into situations where the bias of ordinary language is not appropriate. Neither should it induce us to speak as if metaphysical assumptions were a matter of indifference. They are relevant to the particular forms of experience, being logical tools we use in thinking about that experience in order to live with it in some way or other.

CHAPTER IV

THE INTERFEROMETER AND
THE COUNTER

[It is taken for granted here that the reader is familiar with the experimental basis of atomic physics. It is now over thirty-five years since the first successful inventions in quantum mechanics made their appearance, so there are available good explanations. However, despite the effective learning process by which young physicists acquire competence to use quantum mechanics in their work, there persists, outside the subject proper, a continued interest in philosophical questions associated with it. By their form these questions reveal bias in favour of treating the matter as generally as possible and ignoring physical details as irrelevant. This is the point of view that the discussion to follow is aimed at undermining—one really cannot understand these things if one is determined to have a sweeping simple abstraction, for they are more subtle than some mathematicians and philosophers seem to assume.]

As one of the most used illustrations of the dilemma of classical physics in the face of atomic phenomena, a photon counter is imagined to detect the light in Young's arrangement of two parallel slits in a screen opaque to light. On one side of it there is the source of light intended to produce on the other side the interference pattern consistent with the wave theory of light. The statistic of photon counts registered in the counter agrees on the average with the expected distribution of light intensity, and departures from this are accounted for by the laws of statistical fluctuations. Since the interference pattern can be observed even when the light intensity is so weak that the probability of the simultaneous presence of two photons in the interferometer is negligibly small, we seem forced to conclude that the individual photon 'interferes with itself'.

If we were dealing with golf balls coming through a screen with two holes in it, the probability distribution for the arrival of balls beyond the screen when both holes are open would be the weighted sum of the probability distributions obtained when each hole is

46

separately open and the other closed. The ball passes through either one hole or the other. With this example in mind we are led to say that the classical laws of probability do not apply in the motion of atomic entities when interference occurs. This view of the matter has been challenged* from the point of view that the law of total probability which is appealed to should not be relevant in any case because of the general principle in quantum mechanics that the only meaning to be ascribed to statements concerning position, path, momentum, etc., of an elementary particle, is in terms of the outcome of experimental observations concerning them. For the present we shall leave examining this principle, merely remarking that resorting to it illustrates the pitfalls of the abstract approach to a difficulty in understanding, because it displays no interest in the physicist's intuition about things existing and moving: it is judged sufficient with the aid of the general principle mentioned above, that the logical way is open for the interference of probabilities. This argument in fact misses the point. Under what circumstances are we entitled to say which slit the photon has passed through? The physicist wants to be shown how to imagine what is going on, and no amount of mathematical rectitude will serve in place of this. Indeed, there have been striking instances in the history of mathematics when a logical argument prevented mathematicians from recognising a major mathematical invention presented to them. It is therefore desirable to heed the symptoms that something is unclear, and look at the matter more closely.

Let us devise a means for detecting from which of the slits a photon appears to come. Let us use a telescope capable of resolving the slits and arrange to have two detectors, one on each of the two distinct images. Clearly this arrangement would permit one to associate each detected pulse with a particular slit. Quantum theory does not prevent one from using such an optical device. But in order to show the interference phenomenon we require a

---

* B. O. Koopman, *Proceedings of the VIIth Symposium in Applied Mathematics of the American Mathematical Society* (New York: McGraw-Hill Book Co., 1957).

single detector on which both of the images not merely fall but on which they have overlapping diffraction patterns. To resolve the slits, the aperture of the telescope placed at distance R from them must exceed $2R\lambda/a$, where a is the separation of the slits, and λ the wavelength of the light; to show the interference, the aperture of the telescope must be less than $R\lambda/a$. These conditions can be met with a given telescope only if the slits are resolved when the telescope is close to the slits, and the interference pattern exhibited when the telescope is far removed from them. In the former condition, the aperture of the telescope has to be wide enough to separate the light coming from the two slits and therefore embraces many interference fringes that could be separately observed only with a much less directive receiver. The latter, of course, would not permit one to say from which slit each observed photon had come, and any device introduced at the slits to signal the passage of a photon would, as is well known, spoil the interference pattern.

It seems then the essential point is not that one cannot tell through which slit a photon passes as asserted by the logician, but that any arrangement for getting this information is incapable of concurrently exhibiting the interference phenomenon. What we are to say about combinations of probability regarding photons is thus relative to the kind of detector we have in mind, and depends on whether or not the elements of the divided source of light are coherent, that is, in a well-defined stable phase relation.

Consider a vehicle bearing two tail lights receding from the observer in a straight line. At some point the two lights will appear as one. A corresponding observation could be made of a pair of slits through a telescope with an aperture of variable diameter. When this diameter is large, the slits are seen separated in the field of view of the telescope. As the diameter is reduced there comes a stage when the images become blurred, overlap and become indistinguishable. Of itself, this is not the whole story of interference. The foregoing could be observed whether the radiations from the two slits were coherent or not. What coherence introduces into the phenomenon is the variation of the intensity of the joint image as the telescope is moved about. In particular, if the telescope were placed to view an interference minimum, the superposition of the images would reduce the total intensity of the light received by the telescope. Thus, although the images appear to represent the same source when they are superposed, they represent this source in different ways. The spatial separation of the slits is taken into account in the phase difference that controls the interference, and this phase difference (ϕ) could be deduced (mod 2π) by intensity measurements.*

To show interferences we require coherence and a detector of sufficiently small aperture which is non-directive. One does not always use an optical system with lenses or other directive receivers to examine an interference pattern. With a counter the opening by which the light is admitted is in practice very much larger than the wavelength of the light, but of course much smaller than the distance between adjacent fringes in the interference pattern. The role of this aperture is merely a stop; light from any direction in a very wide angle cone will pass through it. It does not act like a radio antenna array on which the distribution of phase produced on its elements by the incoming radiation determines the strength of the received signal, and for which there is a definite receiving antenna pattern rigidly attached to the physical array.

When the interference pattern is received on a photographic plate the physical effect of the light is on individual grains in the emulsion. The very finest of these are about 10^{-4} cm in diameter, but this is not important in principle here, because the process by which the light is absorbed is an atomic one occurring somewhere in the grain and made visible by chemical development of the whole grain. There is no problem concerning detecting interference.

The interference is connected essentially with coherence in the source, indeed the test for it is the visibility of the fringes. Ever since Zernike's researches, opticians have developed this and ideas of partial coherence†—far removed from the simplicity of the

---

* If I_1, I_2, and I_{12} denote the intensities when the slits are acting separately and jointly then $\cos\phi = (I_{12}-I_1-I_2)/2\sqrt{(I_1 I_2)}$.

† See E. H. Linfoot, *Recent Advances in Optics* (Oxford, 1955).

conceptions we have discussed. One cannot avoid discussing phase in these questions: it is just as important for optics as intensity, and difference in phase can be measured in various ways.

Thus we reach the physically obvious position that the statistics of photons (or of any other particles) with respect to interference are governed by the laws of optics. Since the intensities of super-posed light fields do not add like independent probability densities, classical probability theory is not relevant to that superposition. It is necessary to find a logical place for phase difference.

As is well known, the hypothesis in atomic physics that the phase is the mechanical action divided by $\hbar$ led forty years ago to Schrödinger's equation and initiated the evolution of quantum mechanics into the impressive system we use today. Teachers of the subject have often presented it via the analogy between particle dynamics and optics that inspired Hamilton's work, but usually the formal treatises do not depend on that analogy. Instead, emphasis is laid on the inadequacy of classical conceptions, and on the need to learn a formalism with conceptions about measurement and motion differing greatly from the classical.* It must surely be admitted that the engineering of the theory has been effectively carried out. Nevertheless, it is not hard to find philosophical puzzles connected with it, or thoughtful physicists who are un-comfortable in the face of accepted doctrine as to permissible statements about physical nature at the atomic level.

In thinking about instructing learners of the subject one should surely be concerned to present the mathematical formalism with physical understanding of its appropriateness, not merely to exemplify how it is used. Far from 'the interference of prob-abilities' characteristic of quantum mechanics being an obstacle to understanding how probability is connected with the motion of atomic particles, it should be found to be peculiarly consistent with their atomic nature.

So far we have noted the role of the detector's receiving aperture in determining the visibility of interference fringes, it being assumed that the necessary degree of coherence is exhibited by the

---

* For example, D. Bohm, *Quantum Theory* (New York: Prentice Hall, 1951).

beams of light that are superposed. The detector of the particles which are counted may be regarded in a way that will help us better to understand just what it means not to be able to tell which path the particle has traversed to reach the detector. Consider an interferometer in which a beam of parallel light is split by a half-silvered mirror and which, by means of a second half-silvered mirror, produces interference at the detector of the superposed partial beams that differ in path length and therefore in phase. Such a device is illustrated in figure 1. The pair of full mirrors Y and Z may be mounted on a carriage that may be displaced by means of a very fine screw at right angles to the direction of the

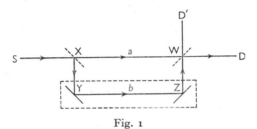

Fig. 1

light passing between them, in order to vary the phase differences between the beams entering the detector D, one through the half-mirror W and the other reflected from its rear surface.

First, we notice that D receives only part of the light passing through the instrument. A second detector placed at D' can also receive interfering beams; the intensity changes produced at D' by interference will be complementary to those observed at D. This particular arrangement thus directs attention to the distribution of intensity of light over all possible emergent channels, whereas with Young's experiment we tend to overlook the fact that interference redistributes the total light energy passed by the pair of slits.

With this interferometer, if one wished to tell whether a photon had traversed path a or path b to the detector, it would be necessary to tilt the mirror W and to displace it, or the detector D, laterally so as to separate the beams. A simple calculation will show that the amount of tilting of W and displacement of D required to discriminate between the paths of the photons regarded as particles

51

subject to the uncertainty principle will destroy the interference at D. This result is of course compatible with the statement made above in optical terms about the receiving aperture of the detector of interference fringes in Young's experiment.

Thus, if we think of an interferometer in terms of the combination of the probabilities for counting particles, we must keep in mind that the particles cannot be distinguished in respect to the path by which they reach the detector without violating the uncertainty principle, and when it is arranged to distinguish the paths in accordance with the principle, the interference is destroyed. Interference is observed when the changes in detector orientation and position, or in the detector's receiving pattern, required to distinguish the classical interfering paths *at the detector* are more discriminating with respect to the particle than the uncertainty principle allows. In this way we see how the separate paths in space, clearly established to be separate by valid physical observations, do not present separable possibilities at the detector consistent with the uncertainty principle. The two paths, which are indeed physically distinct, cannot be distinguished by means of the radiation studied and the interference detector used, without requiring a degree of precision denied by the principle. That there are two or more paths leading to interference is shown by the interference pattern. This is a consideration of great importance to which we shall return. In an interferometer, if the classical law for combining independent probabilities were applied to the transmission of particles to the detector, we should in fact be claiming to escape the uncertainty principle.

Thus by examining the physical situation one may penetrate to a better understanding of the relevance of interference to the atomic nature of the particles and grasp that the wave function represents a great deal more than is required by probability theory.

In demonstrating interference the difference in path length must be small compared with the lengths of the two trains of interfering waves. In particle terms, this is equivalent to surpassing the limit set by the uncertainty principle for spatial resolution of the particle's position in the direction of propagation, and to surpassing the limit of timing resolution permitted by the principle. For the length (l) of the train determines the uncertainty in the wavelength.

If the difference in path length were large enough the two wave trains would not interfere, because, not being superposed at the detector when they arrive there, they could be detected separately. The uncertainty in the phase difference between the two trains is $2\pi \times$ difference in path length$/l$. This uncertainty must be kept small to preserve the phase relationship on which interference depends.

The explanation of interference is simple only for an infinite wave train: with a finite train there will be an advance signal from the shorter route, and later a delayed signal from the longer route. So the duration of the wave train must be very much greater than the difference between the times of passage of the centres of the train by the two routes. For the particle, however, the different times of passage by the different routes in the interferometer are not distinct even though we compute them to be so from the measured lengths of paths and the known wave velocity. These distinctions are not valid for the particle because of the uncertainty principle.

We must distinguish between an experiment planned to demonstrate diffraction or interference well, and one in which the observed pattern would not point to diffraction taking place in a simple clear way. In theory, the diffraction always occurs, but only certain experimental arrangements are favourable to exhibit it clearly. To produce the interference the wave-front must be altered from a simple spherical wave. The physical arrangement transforms the wave-front from the source and alters the amplitude distribution. From a pair of slits illuminated in phase, the equiphase surfaces are approximately oval but have a ripple on them which is the more marked the closer one approaches the slits.* On the source side of the slits there will also be distributions in amplitude and phase resulting from the superposition of the incident waves and the secondary waves from the screen bearing the slits. There is a

* This phenomenon can be exhibited in detail with microwave radio waves.

diffraction phenomenon on both sides of the screen and a reaction of the source due to the reflected waves, but this effect would be small and insignificant for a reflecting flat surface with a distant small source. Nevertheless, it is there in principle. It seems, therefore, that the elementary discussion of interference is much too simple. The wave system is quite complicated and the behaviour of the source in emitting the waves is conditioned by its environment. When the source is considered to emit particles it is assumed that its behaviour is quite independent of its environment. We think classically of particles ejected from a point in a definite direction with a definite speed, or with a definite probability of ejection in a particular direction, irrespective of the spatial environment of the source.

Just as this relation of particles to the source has to be treated in the light of antenna theory, so the phenomena of diffraction require us to depart from the simple classical ideas of particles and the fields of force in which they move.

In interference the particle seems to be presented with the choice between alternatives that are clearly resolved on the basis of other unassailable physical evidence. Yet under the conditions for detecting the particle, these alternatives are not resolvable *at the detector* because of the uncertainty principle. An interference experiment seems designed to split the particle by offering alternative classical paths each of which it seems the particle must occupy simultaneously in order to produce the interference pattern. From the point of view of the atomic entity that cannot be split, these paths cannot be alternative possibilities in the sense of classical physics under the conditions of the interference experiment. The arrangement for showing interference cannot in the very nature of things distinguish by which path the radiation appears to come. This consideration is relevant to the waves as much as to the particles. The wave-front must be divided by the optical arrangement used. Yet the device for detecting the pattern cannot resolve the sections of the divided wave-front. So from this point of view the interference phenomenon is not so incompatible with the particle as is sometimes supposed.

It is well known in optical theory that each maximum in the diffraction pattern due to an array of slits is separated from its neighbours by minima at which the extreme rays from the array differ by one half wavelength. The shift in relative phase is 2π. Thus the range of the diffraction pattern encompassing a single fringe corresponds on the one hand to the limit of resolution set by the uncertainty principle for distinguishing the extreme rays as possible paths of a particle with momentum given by h/λ; on the other hand, it corresponds to the change by the amount h in the range of values of the action along the rays that are the classical paths by which particles might be expected to pass from the grating or array into the observed diffraction pattern.

Thus if we regard h as the indivisible unit of action, one fringe covers classical possibilities that are not distinguishable with respect to action. We should point out also in this connexion that if we are dealing with the fringe system due to a widely separated pair of slits at a great distance from them, so that the width of the fringes is substantially constant across a part of the field in which there are many fringes, we should retain the classical law for combining the probability distributions in the counting of particles, provided that we accepted a minimum length in specifying the width of the aperture of the detector. Under these circumstances the atomicity of action is translated into atomicity of detector aperture width. In the process the interference phenomenon would disappear. In this context, interference appears as subatomic; it seems to present spatial relations for the particle with a refinement surpassing that permitted by the existence of Planck's constant.

In investigating interference, of course, we depend not only on the refinement permitted by the atomic entities that we think of as passing between the source and detector through the interference arrangement. We have available to us the resources of the workshop and the laboratory for making a fine detector slit and for measuring its position accurately as we traverse the interference pattern. We are therefore quite justified in representing the distribution of intensity through the pattern in the classical way. Our problem is to connect the arrival of a particle we count in this

55

arrangement with the emission of a particle from the source. Our classical concept of this connexion by the continuous motion of particle in continuous space and time fails. To compute the probability distribution observed by means of the counter we resort to waves, and accordingly are faced with the metaphysical difficulty regarding the existence of the entity we imagine to leave the source and enter the counter. In order to emphasise this dual aspect of the phenomena, some men wish to speak of wave-particle radiation, and others of the principle of complementarity. These devices merely turn away from the problem of analysing what we are doing in thinking of waves as the connexion between the source and the detector space, for the existence we are thinking about is supported by events we can observe if we arrange to do so.

The origin of an elementary particle is not a creation out of nothing. One might say that the source of the particle exhibits the departure of the particle in the recoil or change in electric charge, angular momentum, magnetic moment, in fact in the change of the source's physical properties. Thus we can connect the particle with the source. What the conceptual apparatus of 'wave-particle' radiation achieves in thought is to connect an emission event with a detection event (or events). These events constitute the information borne by the physical substance of the real apparatus that we use to study the physical process we imagine connects them. Quantum mechanics seems to write finis to the open process of representation by asserting that we cannot go farther; and this is true with respect to the classical method. Yet using quantum mechanical methods, physicists go on inventing models of microphysical systems and performing experiments to test the models and determine numerical parameters relevant to them. As experience evolves, we go on because all physical representation, being based on *a priori* assumptions concerning space and time, provides us with a canvas on which to draw the progressively more refined pictures derived from experiments with high-energy particles. We apply our representation by means of physical instruments and by conventionally established modes of procedure dependent on theory and evolved from experience and judgments about that experience. The fact that we have encountered the paradox about the apparent existence of substantial entities (particles) that are represented as waves shows that somewhere in our thinking we have erred. The paradox in question points to an inconsistency in our use of physical concepts. Quantum mechanical formalism like a ritual seems to regularise our procedure. Of course, it does more than that for it is a highly developed intellectual structure embracing creative elements of great variety, suggestiveness and power in the evolution of physics. The essential difficulty is to reveal why we are confused.

We have noted that we can establish the emission of a particle by a source in an event. We can ascertain events in which the particle is detected. These events are individual. They involve discrete changes in the physical systems with which 'the particle' interacts. In consequence, the particle differs from the objects of everyday experience because under no circumstances can we know of its presence without disturbing the environment of the particle in a physically significant way. Classical mechanics takes for granted the infinite possibility of refining our observation of the motion of objects in space and assumes the existence of operations which are logically necessary for the observing process, but which are nevertheless physically negligible.

We possess a hierarchy of methods of achieving apparently unlimited precision in the spatio-temporal ordering of events and hence appear to be entitled to use Euclidean space and Newtonian time, or Minkowski space-time, as the basic canvas for representation. The assumption of continuity, carrying with it the infinite possibility of precision in our representation, is forced on us for practical reasons—we cannot use a different space-time for every measuring system. As an example of the kind of difficulty involved, let us consider the correlation of a series of photographs of the same subject that have to be combined to produce the complete film as in aerial photography. Suppose that the pictures are taken with different cameras and on photographic plates of different grain sizes. The way one would go about this task would be to measure the plates with a much more refined optical system and match up the

57

pictures on the basis of practical considerations that are known to the photogrammetrist. The important point is the existence of a super-system for correlating the observations. This idea carried to the limit takes us to continuous space-time. It is merely imagined: it is not realised.

For representing classical dynamical processes we adjoin to our continuous space-time a continuous space of momentum and energy, but in microphysics we are limited by the uncertainty principle which restricts the refinement of specifications in the joint space.

In the laboratory the individual observer receives information in a special order determined by the number of observing channels he employs. In each of them the information appears in serial order. The correlation of these channels in time, or by means of any selecting mechanism that the observer chooses to use, generates information in a different form for 'display'—that is, in a form on which physical intuition and mathematical invention can work. Space-time is one essential scheme of display. We try to analyse our experience so as to extract the space-time facts—the events— to which we attach physically interesting numbers from the total information we receive. It is recognised that this process is limited in practice by intrinsic uncertainty—Brownian motion, Heisenberg's Uncertainty Principle and the indivisibility of atoms. Nevertheless, we go ahead adjoining spaces to physical space-time and thus order the information we receive as observers in fact, or that we imagine to be received as if we were observers with apparatus and detecting and recording instruments.

Under certain circumstances there may be some doubt about the separation and independence of the channels through which the information is collected. For example, channels that cannot be resolved because of diffraction must not be treated as independent, in spite of the fact that they really can be resolved when the information to be transmitted does not undergo interference. The wave process secures the remarkable result that, since the waves emitted by a single atomic source must be treated as coherent, and can interfere at the detector, we really cannot dissect the connexion between source and detector into a number of independent channels in the classical way. The atomic process of emitting a photon and transmitting it through an interferometer to a detector really cannot be split up. Nevertheless, the diffraction calculation takes into account the presence of the interferometer and its action on the waves. What is regarded classically as a set of distinguishable particle channels becomes a mere division of a wave-front— a splitting and subsequent reunion through the interference of waves.

Whereas an ordinary physical object can be divided into parts that are also physical objects, an atomic particle cannot be subdivided in this way.

A particle is an atomic entity that cannot be split in space. So it might appear that if the particle emitted by a source exists in space between emission and detection, the multiple channels required in physical space-time for diffraction and interference must somehow or other coalesce or be by-passed in the space in which the particle exists, otherwise the particle cannot exist.

On the other hand, if the detector is set to show interference it cannot distinguish the channels one from another. This suggests that the different channels, as revealed by other means of exploration, may be periodic copies of one channel under the conditions of the interference experiment. One might say the channels are in fact distinct but they cannot be distinguished by the particle detector system which appears to make them fall on one another like the positions of a rotating pointer in successive revolutions.

Here we are tempted to look on space with the kind of ingenuity that enabled Sommerfeld to use the theory of images in a Riemannian space to solve the problem of diffraction of electromagnetic waves by an infinite half-plane bounded by a straight edge. But this approach misses the point. The particle is localised only by the detector. Only channels resolved by the detector are distinct channels for the particle in the classical sense.

Since the particle is atomic it must connect source and detector in an unsplit way. The conditions for producing interference are consistent with this. The moment we have channels that can be resolved by the detector we no longer have interference at that detector of the waves by these separate channels.

Thus the essence of interference taken in connexion with a detector suitable for showing it, is that the multiplicity of channels relevant to the interference arrangement is not resolved by that system; the theory of the propagation of waves is consistent with this, for we consider the field at a point measured by a receiving system with zero directivity.

It seems quite unrealistic to attempt by the fiction of a mathematical operation to make the two or more channels coincide. In the Young interference experiment, for example, we have to depend on the same kind of limitation of resolving power that prevents the distinguishing of the channels by the detector. The two slits must lie in the same channel. This is obviously so from the detector end, for it cannot resolve them; and unless it has the necessary directivity in its pattern of emission, neither can the source select one slit rather than the other.

Where our explanation fails is in tracing photons from the source of light to the detector. For interference it must be impossible to tell which slit the photon traverses. Thus while the wave description shows a wave front impinging on a screen with a pair of slits in it, the particle description requires that passage through one or other of the two slits cannot be resolved and is a matter of indifference for the detector when interference is being observed. In the optical calculation both slits enter the picture equally. In the particle picture we seem forced to say one slit or the other. This is the state of affairs that led Dirac, for example, to speak of the particle as being in two states or as going through both slits. The wave does not encounter the particle difficulty regarding passage through the slits, but it encounters the difficulty of retaining the substantial integrity of the particle's energy and momentum and other properties. Thinking of energy, for example, as a substance localised in space, we seem to have no choice about the substantial existence of the particles in space-time, but whenever we turn our attention to wave properties we encounter insuperable difficulty in imagining how the particle traverses an interferometer to exhibit interference. The particle must not be asked to face as alternative channels those relevant in the optical calculation. The latter seems to represent what is going on at a level of refinement that contradicts the atomic unity of the particle.

While we may be tempted to think that the passage of radiation through a single aperture raises no problem about splitting the particle, whereas passage through a pair of apertures does, we are really deluding ourselves. The diffraction pattern occurs with a single slit. Corresponding to covering one of the pair of slits with a shutter, we could cover part of the single slit or, *nota bene*, we could increase the width of the slit; it would be just as relevant.

In a diffraction process, the wave carries the particle properties intact—so that they are not split. But the information so carried is not like a radio message—a modulation of a carrier—it is intrinsic for the wave itself; a different message requires a different wave. Modulation of the wave bears information regarding physical conditions at the source and in transmission.

The splitting of waves and their superposition as in an interferometer appears to have no particle version except in terms of probability. Now in experimental physics probability has nothing to do with the existence of a single entity and its motion. The particle goes on its way. Probability requires the study of numerous instances that are supposed to be the same in some respects but not in all.

To obtain a probability distribution we observe a large number of individual instances in order to accumulate the statistic. In a simple Markov process the statistical trend evolves from the bias associated with the individual elementary steps. Indeed the former reflects the latter. Consider the random collision of a particle with a small sphere as used classically for the kinetic theory of gases in treating molecular collisions. One considers individual classical orbits under differing initial conditions on the motion, assumes a probability distribution on the occupation of these orbits and then computes the distribution in space of the outgoing particles. The essence of the statistical calculation is the statistical independence of the orbits. In quantum theory, however, there is no simple connexion between the incident motion and the outgoing. Indeed, because of the Heisenberg Uncertainty Principle, one cannot specify the initial motion in the complete particle way associated with the classical picture. A well-defined momentum vector is represented by means of a plane wave and thus the microphysical localisation of the particle is lost. The system of distinct dynamical possibilities for the incident particle is quite different from the classical. Similarly there are restrictions on discriminating the outgoing motions.

Thus we are confronted with the impossibility in quantum physics of exhibiting a probability distribution of the motions of a particle in the limiting form used classically. No matter how we may try to arrange the matter we cannot escape enforced ignorance (as judged from the point of view of the classical continuum) and are unable to produce the singular probability distributions over all relevant variables that correspond to classical certainty. Accordingly, the idea of investigating the output probability distributions for singular input distributions cannot be applied in the classical way to determine the output for a random input. All inputs have some measure of randomness. This should be placed in contrast with the classical idea of specifying a statistic of what happens to particles fed into a stochastic process such as Brownian motion. Classically we can start the particle off in a particular direction with a definite speed from a particular point of space at a certain instant. The statistic results from counting the distribution of the end results when one or more of these initial quantities is subjected to variation according to random input in respect to them. In quantum mechanics the input is a wave system determined as regards amplitude and phase distribution to correspond with some physically appropriate way of bringing the particles into the system under study. For example, we may consider a plane wave incident with a specified wavelength, or we may treat a localised source of particles emitting the waves corresponding to them.

In either case we start off the theoretical consideration of the process by specifying a wave. We are not to imagine the corresponding particle as observed at all until the final detecting arrangement is reached, otherwise the physical system under study will be altered in the process. There is then an injunction against imagining an electron, for example, moving through the spatial continuum that houses the apparatus of the experimenter who is to detect it. That is, the concept of the localisation of a particle without prejudice to its future motion is not permitted. Nevertheless, this is not strictly true. Consider a proton travelling through a gas and leaving a trail of ionised atoms or molecules in producing which only a small fraction of the proton's kinetic energy is used. Then to a certain degree of precision in the description this can be thought of just like a classical motion. It is when we try to carry this form of description to the limit of the continuum that we encounter difficulty. We must not treat an imagined localisation of the particle without any detectable physical effect as equivalent to an imagined localisation with a specific physical effect imagined as having been detected. We have to exclude the assimilation of these two concepts of localisation. Since the detection is possible only if some matter is encountered we may say that the only points of space that *can* be occupied are points at which actual physical interaction with another entity is represented. The wave process seems to represent the possibility of localisation in empty space—i.e. the detecting system is not specified, but the physical process of detection is conceived to occur when the particle is detected. Instead, representation has to be based on particle-emitting events and particle-detecting events (with physical accompaniment in both cases). When we think of the particle having to traverse intervening space-time we are concerned with interposing a new particle-detecting event, but in so doing we should alter the probability of the later detection. The wave process enters to treat probabilities of detection: actual detection implies physical interaction between the radiation and an atomic system. The mere coupling of waves is not enough: we require the actual commitment that the physical interaction has taken place, i.e. some physical system does register the effect and thus 'detects' the particle.

Thus the particle exists in space-time as a localised entity only in the events that mark its creation, its disappearance or its interaction with some other physical system. Consequently when we talk of the probability of the particle being found at a certain place at a certain time as if the particle were there without our looking at it, or as if our looking at it were a matter of physical inconsequence, we are in fact putting the word 'probability' into the wrong context.

In any actual detecting arrangement we are concerned with the probability of detecting the particle by that particular arrangement. This probability depends on the wave in the vicinity of the detector and on the physical process that is detected. So what we are doing in saying that $|\psi(\mathbf{r})|^2 \, dV$ is the probability of the particle existing in element dV at $\mathbf{r}$ is adopting an abstraction. First, we leave the actual detecting process unspecified. Secondly, we assume that whatever detecting process is actually used, the incident wave field will be unperturbed by it, and thirdly, we treat the measurable probability as composed of two factors, the field amplitude squared and the efficiency of the detecting system. This bears a certain analogy to the way in which electric and magnetic fields intervene in the calculation and the measurement of the physical effects due to electricity and magnetism.

The wave field is in effect a statistical invention; it applies to connecting the detectable events in which physical effects actually occur. These are the only physical reality to be observed. They would be there to see if we cared to look. Whereas in the absence of physical effects that change the odds for possibilities not yet realised, there is, strictly speaking, nothing to observe. It has not happened.

The purpose of the somewhat involved discussion in which we have engaged in this chapter has been first to undermine the idea that non-physical treatment of the wave-particle dilemma can be philosophically effective. We cannot adhere to the ordinary idea of a particle in the sense of an ordinary object existing in space continuously, and thereon regard the waves as the statistical apparatus for some kind of stochastic process in space and time by which the particle moves in space from its source to the place where it is detected, and as if the uncertainty principle should be regarded as statistical in origin.

The phenomena that we represent by means of waves in continuous space and time involve atoms. The wave representation seems to set up another kind of existence for the motion of an atom (or elementary particle), but the unity of the wave process with respect to interference shows that this idea is untenable.

Planck's constant h, the atom of action, is an intrinsic element in the physics of the motion of atoms. As the connecting link between the particle and the wave pictures, it governs the scale of physical magnitude where one yields to the other in appropriateness for describing atomic phenomena.

The unity of the particle conforms with our simple idea of an atom for it has no spatial parts. The unity of the wave process required by interference seems not to reflect the atomicity to which it also must be relevant. So it is clear that our concepts of atoms and motion are the basic sources of our intellectual puzzle.

CHAPTER V

MOTION

In physics we are interested to describe changing states of affairs. The theoretical inventions through which these descriptions have been systematically elaborated work by exploiting the resources of an appropriate branch of mathematics. The latter has sometimes indeed been invented and developed in connexion with physics. When we look back to the beginning of mechanics, we find the simplest of these ideas in representing the motion of an ordinary material body by means of a point moving in the Euclidean continuum. This achievement was associated in the foundation of dynamics with the invention of the differential calculus, the concept of instantaneous velocity, and indeed of the instantaneous rate of change of any physical quantity with respect to the time or space, because all change was regarded as continuous. While discontinuous changes in momentum seemed in accord with experience of the collisions of moving bodies, they were interpreted as due to the action of large forces between the colliding bodies during the very brief interval of their contact and mutual deformation. As Hertz pointed out in the Introduction to his *Principles of Mechanics*, our intuition is to regard as continuous the very rapid changes that are represented as instantaneous discontinuous jumps connecting the motion before and after the collision. That is, we accept discontinuous change as a convenient representation of the effect of the rapid change over a time interval, so short that it belongs to an order of magnitude very different from the much larger intervals we should consider appropriate for observing the physical variable before and after the jump. Hertz indicated our metaphysical bias with respect to dynamical processes, namely, that by improving the refinement of observation we can always reveal as continuous what appears on cruder examination to be discontinuous. These views were expressed less than one decade

66

before Planck advanced the hypothesis of quantised mechanical action and before the series of related discoveries that launched atomic and nuclear physics in the twentieth century.

Our approach to classical mechanics, and its far-reaching involvement with continuous representation, more often than not ignores the metaphysical assumptions of ordinary language and our attitude to the physical things on which we depend in experimental observations of mechanical phenomena. The objects whose motion we observe and represent classically can always be recognised by processes that are assumed to be dynamically insignificant. We depend on them to connect the appearance of the object in one event with its appearance in a later one even when the object undergoes physical change. Provided that these auxiliary changes that are not represented in our dynamical description are unrelated to perturbation of the motion under circumstances we judge to be physically relevant, we ignore them. It is assumed, in fact, that labelling by some means or another is always possible to support the application of our physical representation of the motion. Since labelling must be associated with ordinary matter, labels submit with it to the law of continuous transfer in space and time. A label recognised in two distinct events A and B must in our representation be observable at any intervening time at a spatial position lying between the places of A and B on some continuous path passing through them, and as the time-delay of B with respect to A is made smaller, the interpolated event will tend to occupy a position on the straight line joining the spatial points marked A and B between them.

Thomson's discovery of the electron and Einstein's invention of the photon introduced into physics objects that certainly cannot be labelled in the classical way. Nevertheless, under the influence of the classical tradition, and the habits of speech associated with it, until the invention of quantum mechanics, physicists retained in their thinking the disposition to treat the motion of electrons and photons as if they could be labelled, and as if their motion were continuous.

The invention of cinematography, and the diverse apparatus for presenting observations on microscopic time-scales with the aid of the cathode-ray tube, change our attitude to these simple metaphysical assumptions about motion. We know that although motion on the cinema screen appears continuous it is in fact not so. It is presented by a succession of static pictures each frame separated from the next by darkness. By means of it, we easily apprehend that a discontinuous motion could appear continuous when observed within adequate resolution. We also see that discontinuous motion involves the processes required to switch off the picture at each frame, and switch on at the next. If an object appeared naturally in this way, we should say it is created in one event, endures at the corresponding place, and then is annihilated there, the whole process being repeated at another place and so on. Our inclination is to think only of the visible object when it appears, and to fail to recognise that if we are to speak of the reappearance of the object, the connexion between appearances is just as important. How do we judge that it is the same object that reappears? All that we have is the repetition of an appearance.

For the purpose of discussing this matter we shall find it convenient to examine a much simpler phenomenon. Consider a row of lamps that can be switched on and off individually according to any programme we care to adopt. Let us think of switching on the lights in succession along the row, and switching off each light before switching on its neighbour. Viewed crudely, the appearance of the row of lamps can be looked on as the continuous motion of a bright spot along it. On more detailed inspection, the dark interval between successive appearances of the light and the spatial separation of the lamps are evident. At a particular lamp two processes occur, namely, switching on and switching off the electric current through it. After a certain time interval this is repeated at the next lamp, and so on.

We might represent the events on an xt-graph, x representing distance along the row and t representing time. The row of dots representing switching off is displaced to a time slightly later than the row of crosses representing switching on. In our diagram (figure 2) the successive events are joined by straight line segments

68

of a stepped line. Seeing the row of lamps as a continuously moving spot of light corresponds to ignoring the detailed structure presented in our graph, and seeing the line as it would appear as reproduced by a crude printing process. Since the lamps are distinct and separate, the straight line segment joining a dot to the immediately succeeding cross does not represent spatial or temporal possibility for the appearance of the spot of light. There are no intervening lamps. Likewise there is no means provided for

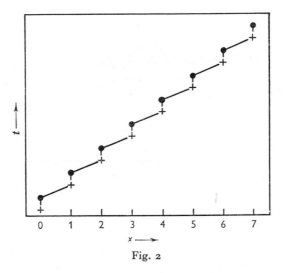

Fig. 2

showing the light at times between switching off one lamp and switching on the next. Thus our representation in a diagram brings to our attention certain aspects of the continuous representation that are irrelevant to the discontinuous and discrete. The successive steps by which we pass along the x-axis are identical and they each involve displacement also in time. We might think of this as reflecting the speed of propagation of a signal from the lamp being switched off to the next one which is to be switched on. But unless we know of an actual process of propagation which we can investigate, this way of looking at the matter is inappropriate, and indeed, if we arrange to switch the lamps on and off by means of a timing device it would be quite incorrect physically.

The remark that has just been made brings to our attention that the real physical connexion between the switching off at one lamp and the switching on at the next is not representable on our graph. So far as our picture is concerned, it is an atomic connexion. We cannot get inside it in our representation, except in the irrelevant way presented by the continuous two-dimensional space in which the graph is drawn. On the other hand, the connexion between switching on and switching off the same lamp is meaningful. The graph presents an interval of time during which the light can be observed there with whatever refinement of timing we care to adopt in observing it.

Substantially, the foregoing considerations would apply to the appearance of the moving spot and our representation of it if we introduced variations in the successive time-intervals between the lighting of adjacent lamps. We could even consider a family of graphs that differ from each other due to statistical fluctuations of the timing-mechanism that controls the appearance of the light. But all of these would fall essentially under classical representation if viewed crudely.

Let us now consider a different use of the row of lamps. Let us imagine that after the lighting of the first lamp the lighting of the next $(n-2)$ lamps is suppressed and the nth lamp alone is lighted. Imagine that this is observed many times, and that n is subject to random variation. How should we interpret our observation? We should soon note that after the lighting of the first lamp, the next appearance of the spot of light occurred at a later time dependent on the position of its appearance in the row. We should then be tempted to think that the appearance of the first light is followed by the emission of something that we cannot see, propagated at a certain speed, and that it causes the light to go on in one of the lamps by chance. But this theory would not work physically under our assumption that n is randomly chosen, unless we introduced a hypothesis about the probability of lighting a lamp to imitate the attenuation that should take place as the signal passes along the row. If, however, the row of lamps is very long, so that the chance of lighting a particular lamp is small, we should, in treating a portion of the row, be justified in ignoring attenuation. This is a matter of little importance in relation to the main point of considering this arrangement.

So far as our representation is concerned, the connexions between the lighting of the first lamp of the row and other lamps are still atomic connexions. The system does not present intervening appearances of the light spot. Nevertheless, because of our observation of the large number of appearances, and the law connecting the time and place of their occurrence, we are ready to think of something moving along the row to connect the lighting of the first lamp with that of another. What is more, we introduce statistics of detection to account for our not seeing the light appear at intervening lamps.

In establishing the connexion, not only do we imagine something moving but we have to imagine motion according to a particular law that fits the facts. We are the more likely to establish this connexion when the law of motion is a simple one. Since we know of no cause that could change the velocity, we assume it to be constant. If this kinematic law did not fit the facts about the observed events, we might try to fit them to another law just as Kepler did in finding the laws of planetary motion. However, since any law other than that of rectilinear motion at constant speed would require the motion to take place under the action of a force, we should then be required to find the physical cause of the force. We should therefore examine the objects and processes in the environment. If it were found that by changing each of the objects or arrangements of them we produced no effect, or if we found no other object subject to the same force in its motion, then the hypothesis would fail. We should say that the connexion we imagined does not exist. The discovery of the electron depended on processes somewhat like those just indicated. The deflexion of the electron beam by a known electric or magnetic force conformed with the idea of a particle with a definite charge/mass ratio. It is noteworthy, however, that conventional accounts of this and other discoveries in physics usually omit referring to the possibility of metaphysical uncertainty about the existence of the unseen particle travelling between the cathode and anode in an evacuated discharge tube. The metaphysical decision has already been taken. By presenting the phenomena in connexion with measuring the properties of the particles, existence is automatically taken for granted, and, of course, in the context of instructing learners about what has been discovered, introducing the metaphysical question serves no practical purpose; rather, it distracts attention from what should be learned. Explanation of the theory must proceed on the basis of some metaphysical presumption. Indeed, we should say that the theory explained is a theory relative to that metaphysical datum. Does this not illuminate how it is that metaphysical discussion forms no part of the formal account of physics?

Let us now imagine that we have a large two-dimensional array of lamps and that we select for our starting signal a particular lamp in the centre of the array. Let us arrange to light one lamp at a time subsequent to the lighting of the starting signal consistent with the hypothesis that a signal has passed at a constant speed in a straight line from the latter to the former. Let us also arrange that the frequency distribution of lighting over the array conforms with the geometrical spreading and the attenuation we referred to earlier. We should then be in the position after observing a large number of repetitions to infer the emission of something from the signal lamp with constant speed but randomly directed, which has a certain probability of lighting one lamp. The lighting of the signal lamp would then be regarded as the event that marks the departure of the entity we imagine, and the lighting of the distant lamp as the event in which the entity is detected. Alternatively, we may think of the former event as marking the creation of the entity and the latter as marking its annihilation. Our readiness to establish a connexion between the signal lamp and the other by means of an entity moving between them is not unreasonable. It is merely remarkable. We make use of similar hypotheses not only in everyday life but also in science. We see a small boat pass under a bridge and see the same boat emerge on the scantiest of evidence. In connecting the evidence of one set of observations of the stars with a later one we apply a very elaborate theoretical apparatus, the use of which is justified only by plausible reasoning.

72

It is important to grasp that the justification of whatever hypothesis we may be disposed to try does not lie in a logical compulsion to accept it. We acknowledge the limited scope of our view in examining our surroundings, and that we cannot look everywhere all of the time. So our observation is necessarily a sampling process. From these samples we formulate expectations of what we will observe in the future. There is no unique recipe for doing this. We improvise on the basis of our talents, training, skill, imagination and experience and we go on rejecting unsuccessful attempts until we find some measure of success in prediction and gradually establish confidence in our ideas and our procedures in applying them.

The formal treatment of motion in mathematical terms tends to ignore the practical invention and metaphysical manoeuvring through which mathematical representation is applied. These complications are avoided in the mathematical theory which is based on axioms that eliminate from consideration possibilities that can be invented by processes resembling our thinking about the moving spot of light. Let us return to consider the hypothesis that the signal lamp radiates entities to light the lamps distant from it.

We are immediately confronted with a number of questions that arise quite naturally from our experience of learning about our physical environment. We might endow the entity with the property of extinguishing the lamp it leaves and of lighting the lamp it is absorbed in, and speculate that it is composed of a substance that cannot be seen except with the aid of a lamp. What usual physical properties are we to attribute to its substance? Quite clearly on the basis of the information to hand about them we have no means of attaching meaningful numbers to the events which mark the departure and arrival of the entity, so we have no means of attributing properties to them. Suppose, however, that we introduced the spaces of light intensity and colour associated with each of the lamps, so that an emitting event is marked by a flash of a definite intensity and/or a particular colour. Correspondingly we can imagine like distinctions in the lighting of the receiving lamp. By measurement we should now be able to investigate how the entities associated with the transfer of coloured signals

73

of particular intensity move and establish connexions between velocity and colour and intensity. Our further thinking would be governed very much by the particular forms that we found consistent with our observations.

In mechanics of course, we do not admit colour of light or its intensity as dynamically relevant when used in the way we have imagined above. They do not affect the motion physically, they are only connected with it, being transported according to the kinematic law that contains these variables as parameters. If we had two signal lamps, which emitted signals at times appropriate to the simultaneous arrival of the entities at a receiving lamp, which lighted with twice the intensity appropriate to the colour, we should presumably have much better grounds for confidence in the existence of the entities we have imagined. Nevertheless, this does not confirm the decision: that lies in our willingness to accept the hypothesis and make good through further invention whatever deficiencies are revealed by experience.

Just as soon as we imagine the entities endowed with mass we must imagine them as transporting energy and momentum. We should thereon set out experimentally to exhibit this and to measure their mass. Guided by our experience of the particles of physics we could readily invent scattering and other processes by which to investigate their physical properties.

The foregoing discussion was introduced for the purpose of exhibiting how imagination and observation interact to evolve a new concept, and to bring to light the atomic connexions between appearances in a discontinuous motion.

Let us now imagine a somewhat different apparatus behind the scenes of a discontinuous motion. Imagine a pulse of electrons emitted with substantial kinetic energy at an angle to a strong magnetic field in a vacuum so that they describe a helical path. Let us place in a plane containing the axis of the helix a strip of material so thin that its stopping power is negligible; it is capable of emitting light under electron bombardment. We are going to think of the motion of the spot of light on the strip. Here we do have a continuous connexion between the events that are regarded

74

as constituting a discontinuous motion on the strip. It is achieved by passing out of the space in which the discontinuous motion is represented. On account of the artificial circumstances of this arrangement, it serves merely to illustrate that a discontinuous motion may be regarded as the intersection of the space in which it is imagined to take place with a continuous track in a space of a higher number of dimensions. Indeed, we are quite free to adopt this method in order to treat discontinuity by means of continuous representation. Whether this device is significant for physical investigation as more than a representational aid in calculation will appear only in the course of its use. If it leads to new, successful experiments it will most likely affect physical concepts.

We have been discussing the imagined motion of objects we cannot see and have noted how we depend on the forms of classical mechanics to provide rules for connecting the observed events. We may be really prevented from interposing observations at intervening events. That is, every attempt in practice fails. On the basis of this experience are we to say that we cannot interpose an observation, that we have to accept this metaphysical attitude, or do we continue to leave the matter open, saying that some technical inventions or discoveries in the future will enable the interpolation to be achieved? Clearly there is no logical compulsion about the matter. We are free to choose. In choosing, however, we commit ourselves and have to accept the consequences of our choice in doing physics. So long as we insist on the possibility of interpolation we are committed to the metaphysical idea that the object exists in space and moves in the classical way, or at least in some way that can be crudely described as classical, say, in a series of very small discontinuous jumps according to a statistical law under which the appropriate asymptotic trend is established. If, on the other hand, we are ready to accept our experience to date as good evidence that our further efforts to interpolate observations will be unsuccessful, we adopt this as part of our formal thinking and resign from further practical effort to achieve it.

The process which we have just indicated was exemplified in connexion with the special theory of relativity at the end of last century and the beginning of this. A whole family of experiments was performed with the object of finding effects of motion 'with respect to the aether'—the material supposed to support the propagation of light and electromagnetic effects.* The results of the experiments agreed in denying the effects to be expected on the basis of Newtonian relativity, so the hypothesis of an aether in that sense was abandoned and gradually physical theory was committed to Einstein's view.

We should compare these matters in physics with the experience of mathematicians in adapting themselves to problems about construction or proof. The trisection of an angle by means of compass and ruler is the best known example of the former—mathematical proof terminated attempts by professional mathematicians to accomplish the construction. The latter is exemplified *par excellence* in Godel's theorem. In both contexts, attitude of mind is the important element in approaching the difficulties met in trying to do what we imagined as possible, because it determines how we go on. Either we accept rebuff, or we continue to be frustrated. Since the evolution of physics must depend on the efforts of many men in different countries, one or other of these attitudes will prevail in practice and the subsequent development of the formal structure of physical theory will conform with that attitude, until sufficient evidence has been recorded to demand either fresh invention, or that we go back and choose the alternative formerly rejected and accordingly rewrite the formal theory. For practical reasons the latter is less and less likely to happen as time goes on, because a dilemma in thinking appears as contemporary not as the recurrence of an old one.

If we wish to incorporate into our formal thinking that certain possibilities of our former representation are not to be regarded as possibilities in fact, we have to change the representation by adopting a rule, or rules, the effect of which is to exclude these possibilities, or we may invent a new representation which does not present these possibilities at all. We shall remark here in passing,

* See E. T. Whittaker, *A History of Theories of Aether and Electricity*, p. 29 (Edinburgh: Thos. Nelson and Sons Ltd, 1953).

and return later in this book to discuss how it does so, that quantum mechanics uses the second method—in spite of the appearance that essentially rules are superposed on classical mechanics—for these rules govern the logic of the system. They are not mere rules of interpretation.

Now, if we have to deal with connexions between events under circumstances when we cannot interpolate (logical prohibition), in what sense are we justified in speaking of motion at all? Have we not changed our concept of motion? Do we not also in the process destroy the symmetry of the relativity of motion? We can always introduce frames of reference in relative motion in the classical continuum and depending on objects that can be treated classically, apply the transformations which connect physical descriptions with respect to different frames. But to attach a frame of reference to an imagined object between the events in which it can be observed seems unreasonable. This is a point of some philosophical interest for it draws to our attention that the formal requirements of Lorentz covariance in quantum field theory are in fact classical in origin. They apply to symbols that appear in the mathematical elaboration of the theory and guarantee that measurements as represented by theory will be consistent with the Lorentz transformation, etc., in the context of observation in the laboratory, or the observatory, or wherever we place our instruments for the physical observations.

From the discussion of discontinuity in representing motion in the foregoing sense, we turn now to the sudden changes in direction and speed of motion conceived classically that we associate with collisions. We may note that the discontinuous change in momentum could be represented in momentum space as atomic connexions between points in that space, just as we did in discussing the spot of light on the row of lamps. If we are disposed to regard the representative point as moving continuously, so that we overlook the fine detail of momentum change, we connect the values of the momentum by means of the hypothesis of a force acting to produce change. We should be required to look for the physical cause of the force. When we fix our attention on discontinuous changes in momentum, however, we are not satisfied with the crude representation and look for the means by which momentum is exchanged in collisions with the object we observe. Essentially this situation is presented in the study of Brownian Motion. The impulses are communicated to the gamboge particle by the molecules of the liquid in which it is suspended.

Imagine a spot to move on a plane so that it occupies successive positions at instants of a discrete series according to the rule that the displacement in an individual step of the random walk has a well-defined value (a vector) and well-defined variance (mean square distance per step), both independent of position on the plane. To describe motion in this stochastic process we must introduce the successive probability distributions for the appearance of the spot on the plane as successive steps are made. After a large number (N) of steps, the central limit theorem in statistics informs us the probability distribution will have a mean position agreeing with that expected for N iterations of the mean vector for a single displacement. The mean square distance from the average position will be N times the variance for a single step. Relatively, then, the distribution after N steps is less spread out than appears for a small number of steps, for the ratio of the square root of the variance to the mean displacement after N steps varies as the reciprocal of $N^{\frac{1}{2}}$. As the process goes on, the direction in which it proceeds becomes progressively better defined, and from the asymptotic trend we see revealed the mean displacement of a single step.

We may simulate this process in an approximate continuous representation, regarding the probability density distribution as a continuous function of the time. So the time-scale of our description is so large that we treat the change in the distribution in a single step as infinitesimal. If τ is the time-interval per step, $\mathbf{d}$ the mean vector displacement, and σ the mean square displacement parallel to each of the coordinate axes, then $\phi(x, y, z, t)$, the probability density, satisfies the equation

$$\tau \frac{\partial \phi}{\partial t} + \mathbf{d} . \nabla \phi = \tfrac{1}{2}\sigma \nabla^2 \phi.$$

The solution of this equation that represents the process we have been discussing is

$$\phi = \left(\frac{\tau}{2\pi\sigma t}\right)^{\frac{3}{2}} e^{-\tau|\mathbf{r}-\mathrm{d}t/\tau|^2/2\sigma t}.$$

When $t = N\tau$, this form exhibits that the variance after N steps is proportional to N. When d vanishes, ϕ satisfies the diffusion equation which bears a formal resemblance to Schrödinger's equation for a free particle of mass m

$$\frac{\hbar}{i} \frac{\partial \psi}{\partial t} = \frac{\hbar^2}{2m} \nabla^2 \psi.$$

Indeed, taking $\sigma = (h/mc)^2$, and making $\tau = h/imc$, our expression for ϕ is translated into the Green's function for the wave equation.* The parallel has been noted by a number of writers, but although it has served formal computational purposes, it has shed no light on the understanding of non-relativistic quantum mechanics, for we cannot span a real time interval by imaginary steps.

Nevertheless, extending the domain of physical variables from that of real numbers to complex numbers has served physical representation well in a very great variety of applications. It facilitates the use of mathematical analysis, brings to bear its powerful formal methods in computing, and suggests on many occasions the proper formal scope of physical concepts. It permits us to subsume under a single form the oscillating and overdamped motion of an oscillator subject to variable damping, for example, and it brings together propagated and evanescent modes in representing waves.

In classical mechanics representation by complex numbers always appears as an analytical device. The physical variables we measure take only real values, indeed, this requirement defines the region in the space of the configurational coordinates accessible to the dynamical system with a given energy, because on the boundary of this region the kinetic energy is zero and outside it would be assigned a negative value through the energy equation. Penetration

$$* \quad \psi(\mathbf{r}, t; \mathbf{r}_0, t_0) = \left\{\frac{m}{2\pi i\hbar(t-t_0)}\right\}^{\frac{3}{2}} \exp\left\{-\frac{im}{2\hbar} \frac{|\mathbf{r}-\mathbf{r}_0|^2}{(t-t_0)}\right\}.$$

79

into classically inaccessible regions is possible for the particles of quantum physics. It is compatible with wave conceptions because of evanescent waves with complex wave-vectors which have been a well-established part of electromagnetic optics since last century. By this means imaginary values for the momentum of a particle derived mathematically can be interpreted through the optical analogy. But the magnitude of the imaginary momentum component must be given statistical significance, and this is done in quantum mechanics.

In elementary terms we may regard Schrödinger's equation as amalgamating dynamics and statistics by extending the possible values of dynamical variables to the complex domain. If the wave function ψ is represented by $e^{iS/\hbar}$, where S is the complex number $S_1 + iS_2$, S_1 is the mechanical action, $e^{-2S_2/\hbar}$ represents the probability density. So the imaginary momentum is connected with the gradient of the modulus of the wave function through $\hbar$. As is well known,* when the gradient of S_2 and its divergence may be disregarded in an approximate treatment that effectively replaces waves by the corresponding rays, Schrödinger's equation is equivalent to the pair of equations, one of which is the Hamilton–Jacobi equation of classical mechanics, and the other is the equation which must be satisfied by the probability density over the system of particle orbits defined by S_1 to conform with the particle motion.

Thus by admitting the formal possibility of extending the field of conventional dynamical magnitudes to complex numbers, we create the means for associating the probability distribution with the dynamical field in an intrinsic way. Further, we admit the possibility of a particle appearing in a place inaccessible to it by continuous motion according to classical mechanics and describe such appearances statistically. Of course, Planck's constant is required to fix the spatial scale of the transition region between the classically accessible and inaccessible parts of space or states of

---

* The excellent recent discussion by M. Kline, pp. 3–31, *Electromagnetic Waves*, ed. R. E. Langer (University of Wisconsin Press: Madison, 1962), of the solution of time harmonic problems in asymptotic series clarifies understanding how geometrical optics is related to electromagnetic theory, and classical mechanics to wave mechanics.

motion. Indeed, Planck's constant which first appeared historically as the atom of mechanical action has a complementary role in relation to probability distributions associated with dynamical fields at the atomic level.

In our discussion of motion, we have emphasised the connexion between separate events. In the representation of continuous motion we tend to lose sight of this because we have learned to rely on the derivative of displacement with respect to the time. Being the limit, as the time-interval tends to zero, of the average rate of displacement over that time-interval, it corresponds to zero displacement, so the displacement seems to disappear and takes with it the corresponding need for a connexion between separate events on the trajectory. The connexion we should have in mind is not merely the kinematic connexion by the path we represent in a map or graph. That does not exhibit our dependence on recognising, naming, and other processes by which we would apply the representation, picking out individual events that are occupied by the object whose motion we are observing, or that are imagined to be connected by the moving object. Because of the great variety of these processes from which the experimenter must choose in observing motion, and in using them must be ready to improvise, this aspect of physical representation is not formalised in the mathematical theory. Instead, we concentrate our attention on the transformation of coordinates as the connexion between the events that are significant in respect to the motion. Since in classical mechanics motion is continuous, the transformations must belong to a continuous family. The transformation in a particular instance is represented by a mathematical operation that converts the coordinates of one event into those of a later one on the path. Such an operation could be applied to any point of space at the earlier time to transform it in general into another point at a later time. Accordingly, the transformation connects pairs of events, and seems to serve in the logical place of the physical connexion between those events occupied in the course of the motion.

In dynamics, unless we have to deal with rectilinear motion at constant speed or rotation about a fixed axis at constant angular speed, the naïve approach to transformation introduced above does not serve to carry our thinking very far. It is the idea of a dynamical system and of families of motions or trajectories covering at least part of space continuously that leads to the effective ideas of Hamilton's transformation theory, which connects particle dynamics with ray optics. Historically this theory played an important supporting role in the invention of quantum mechanics, indeed it presents mechanics in a form that facilitates passage from the continuous to the discrete and discontinuous.*

One readily concedes that the evolutionary descent of quantum theoretical forms from classical ones corresponds formally with the practical reality that we depend on classical representation whenever the degree of detail involved is not too refined. We have established a theoretical bridge from quantum to classical mechanics, but there remains the question 'Is this bridge physically relevant?'. Answering this question must introduce a wider view than is needed to establish only the mathematical connexion, especially since the latter ignores completely the inappropriateness of classical forms of speech concerning the existence of the physical entities being discussed.

Schiff's† treatment of the motion of a particle that interacts with a pair of atoms, and his calculation of the probability distribution over the momentum of the emerging particle under conditions conformable to classical approximation, is the most impressive explanation known to the writer. In this calculation, however, the atoms are treated as given in space and their existence is treated in the same classical way in which we regard the diffracting systems we use to analyse radiation.

However, we are not concerned here with these well worked out mathematical systems: we return instead to the idea of uniform translation and the possibility of applying a transformation to the coordinates in our map so that in the transformed map the point formerly moving is brought to rest. Corresponding to discon-

---

* See *On Understanding Physics*, p. 127.
† L. I. Schiff, *Quantum Mechanics*, 2nd ed., pp. 209–13 (New York: McGraw-Hill Book Company, Inc., 1955).

tinuous motion before the transformation we have now to present a discrete series of instants of time associated with the fixed point of space. Here we require the physical connexion between the separate instants: it lies outside our representation. Accordingly, when we represent a continuously moving point in the appropriate transformed system of coordinates, we see no need of the connexion. Something occupies a fixed position. Yet the something must be recognised at different times in order to test the truth of this in applying our representation.

The purpose of the foregoing comments is not to cast doubt on the procedures we use intuitively in doing physics, in particular to represent motion. They serve to show how mathematical formality can usurp attention and distract us from the essential physical informalities on which we really depend, and which are relevant to philosophical discussion for the purpose of improving our understanding.

The transformation inverse to that discussed in relation to uniform translation restores its motion to the point we brought to rest. Thus we think of motion relative to a system of coordinates. In applying this representation we depend on connecting the system of coordinates with the actual world. There may be some physically suitable objects which are represented in the map as occupying fixed positions, in which case we regard the motion as relative to these objects or the system of coordinates attached to them. But this need not be the only way of attaching the coordinate system to reality: we might require that in the representation associated with the coordinate system certain selected objects move in a particular way, as is actually done in astronomy. So it comes about that we are quite ready on the assumption of Newtonian relativity to imagine motion produced merely by changing the coordinate system to one moving with respect to the old. We should notice, however, that if we are concerned with the physical effects associated with motion, it is essential that change of coordinate system itself need not introduce them. Thus if we are considering the interaction of the moving object A with a system E, both of which are at rest in the coordinate system S, we cannot generate the relative motion of A and E by changing S to another system S' in motion with respect to S. This is a consideration we do well to ponder, for the physical definition of a coordinate system is usually taken for granted in our theories. We ignore the physical things to which it must be related in practice; we ignore E, change from S to S', to set A in motion, and then introduce E at rest in S'. We cannot abolish the relative motion of A and E merely by changing the coordinate system in a simple way. But we may alter the relative velocity as indeed happens when we apply the Lorentz transformation to represent the uniform translation of our coordinate system of reference according to the theory of special relativity.

If the coordinate system is attached to objects that are physically relevant to the existence of the entity whose motion we are describing, mere transformation of space and time coordinates is not enough. This is why we have been compelled to introduce into physics the corresponding transformations of the physical fields by which interactions between the physical objects are represented.

In a classical motion the configurational coordinates being continuous functions of the time, the properties that are conserved in the motion are thought of as transferred like substance in the motion, for example, kinetic energy, momentum or rest mass. Provided we can connect the discontinuous motion with a continuous one, we are able to measure these quantities and associate them with the discontinuous motion, even although the kinematic definitions of them are no longer applicable. The connexion may be the interaction we call a collision, or it may be through an effective classical representation of the action of a field of force on the motion, as in the deflexion of a beam of electrons. Nevertheless, in discontinuous motion we have to think of the dynamical properties as transferred in a discontinuous jump, and since we depend on classical supports in using this conception we should recognise that the localisation of the events that are connected by the transfer cannot be precise on a microphysical scale.

CHAPTER VI

ATOMIC EXISTENCE

The basic metaphysical questions that come to mind in reflecting about quantum physics refer to the relation to space and time of the microphysical objects by means of which we interpret our experience. How far is the atomic concept compatible with continuous space-time?

In classical terms we have a manageable concept of physical existence. Substantially we rely on the metaphysical bias of ordinary language, which in turn derives from men's experience over many generations, and the training that accompanies learning the language and its use. Experience with material objects we can see and handle leads us to commit ourselves to continuous representation and continuous motion. So accustomed do we become to this commitment that it is not merely taken for granted; it becomes a logical requirement in our thinking about matter and motion, and the basis of classical representation in physics. We represent things in a space-time map, attaching to the places they occupy the appropriate names or signs that stand for the names in accordance with a legend or table that is used in applying the map. We do not give a table for the names in ordinary language; when necessary, a dictionary aids the reader of the map. Essentially the names have to be learned through use, and likewise the processes on which we rely to use them, for we have to recognise the object to which the name is applied. When we have to deal with a number of like objects which we wish to distinguish, we have to label them by some mark or other which remains attached to the object as it participates in the processes we are watching.

Applying a representation in a map depends on using some of the names in it to establish its correspondence with the reality it represents. Just as we depend on surveying to determine places in the map to which particular names are to be attached, so we depend on measurement to find the place in ordinary space that corresponds to a point on the map. These procedures for recognising, naming, labelling and surveying are not represented in the map, although they could be. Whenever there is doubt about their application, they are so represented in order to clarify representation.

To represent things changing with time we need either a changing map or a spatial representation with an additional dimension.

In physics, as the techniques of measurement have evolved with the growth of physical understanding, the precision of representation has sharpened and, until the end of last century, there seemed no reason to question the procedures by which many varied continuous representations were made and applied. Today, so long as the precision in representation does not approach the limits set by the uncertainty principle, all of the apparatus and procedure in making and applying ordinary spatial representation can be used. We have confidence in ordinary language and rely on the training of physicists needed for applying it successfully.

In observing the motion of a small object such as a particle of dust, which affords no obvious features, and does not easily permit labelling that can be seen, we depend on the hypothesis that its motion is continuous to support the presumption that it is the same particle that we observe through the motion. Whenever the dust particle we have been following comes into the vicinity of another, our ability to follow the motion through the near-collision depends on how well we can resolve the images of the particles. If there is doubt, the identification of the particle we were following terminates, unless there is some other connexion on which we can rely to bridge the gap in the optical observation. This way of using connexion by continuous motion is only one way; there is an endless variety of alternatives that apply physics, just as stars are recognised in astronomy by their position in the heavens at a certain time. Of course, the astronomer is not dependent only on position to recognise a star; its spectrum or regular changes in the brightness of its image serve to confirm identification when they are relevant.

86

The objects of microphysics present to us possibilities like those we have just mentioned, provided that the representation is not so refined that we are prevented from applying classical laws. In interpreting the radioactive processes recorded in a photographic emulsion due to the entry of a high-energy particle, we are ready to interpret individual developed grains 'along a curve' as the track of an electron, and the interpretation succeeds or fails depending on its consistency with other observations with respect to the phenomenon which itself must be recognised and of which other instances have been examined. This appearance of an electron that we imagine to exist throughout the motion and to suffer the ionising collisions which render grains in the emulsion developable is no different in principle, at the level of classical ideas, from the appearances of objects that disappear from view and then reappear. We judge on the basis of plausible reasoning that the object persists and continues to move between the places and times at which we see it. That is, representing the motion on a space-time map by a continuous line through the events in which the object is seen is applied in the following way: each point of the line stands for an event occupied by the object and we could observe it by some means or another if we were sufficiently ingenious to do so. For example, an object hidden in a tunnel could be detected by X-rays or γ-rays or neutrons; indeed this principle has been applied for numerous industrial purposes. This remark ought to warn one about the danger of assuming that something 'cannot be observed'. We must distinguish physical impossibility from logical impossibility. The former is to be judged relative to the techniques available. The latter expresses a commitment to a particular system in thinking.

By classical methods, the charge and mass of the electron were measured; it is the atom of electricity. All electric charges are positive or negative integral multiples of it, but this fact does not embarrass our thinking. We use it as a rule to limit the application of the continua by which electrical phenomena are represented classically. Electrons exist like other objects. We represent them in a map: think of Millikan's oil-drop experiment. The drop of oil captures an electron, acquires its charge and thereon suffers a changed acceleration in the vertical electric field acting on it. At the level of simple classical representation we seem to have no difficulty about its existence in space and time. Yet we do not see the electron.

Since we think of electricity as a substance, the electron imports substance to the oil-drop when it is captured. We have no doubt about the permanence of the oil-drop or of the appropriateness of representing its motion according to Newton's Laws. It is this object that supports the existence of the electron as the existence in the sense of ordinary language. The electron is localised at the drop and is transported in a continuous motion (ignoring the effects of the Brownian motion of the drop).

In spite of the fact that classical mechanics failed to encompass the motion of the electron about the nucleus, even of a hydrogen atom, there never was any doubt that the electron existed localised at the atom of which it forms a part. The atom of chemical substance is an ordinary object so far as classical representation is concerned. It is well to note, however, that the grounds on which we name the atom are not necessarily simple, any more than those we rely on to attach names to objects seen through a microscope or a telescope.

An electron captured from the K-shell of an atom by its nucleus is not thought to exist in the nucleus although it has given up energy and charge to it. And whereas we think of the electron ejected from an atom by light as having existed in the atom before it appears outside, we do not think of a β-particle (electron) emitted by a radioactive nucleus as having existed before that event in the nucleus. The reasons for this change in attitude are found in the physical theories we have committed ourselves to in thinking about the atom and the nucleus. In the extra-nuclear capture of an electron by a positive ion, the energy that is released in the capture is only a small fraction of the rest-energy of the electron that is given up in nuclear capture. Besides, the charge of the electron continues to be associated with its mass in an atom—as revealed by the Zeeman, Faraday and Stark effects, whereas in a nucleus the charge is associated with the transformation of a proton into a neutron. Thus we see how existence is related not merely to location in space but also to connexions between places in other spaces (mass, charge, for example) with each other and with space-time.

When an electron is detected in a counter, on what grounds are we willing to accept the triggering pulses that actuate the electronic recorder as signifying events in the crystal or gas counter due to the electrons we wish to count? We observe the counting rate in the absence of the source of electrons and compare it with that when it is operating in the experimental arrangement. But the basis of our confidence in connecting the emission of electrons from the source with the events counted in the detector is that all physical causes that we imagine could possibly operate the counter—including electronic failure in its circuitry—are ruled out by subsidiary evidence that the experimenter gathers in examining how the experimental arrangement functions when the conditions of operation are varied in different ways. The operations carried out by the experimenter can be described in essentially classical terms— opening shutters, placing absorbers in the path of the electron beam, deflecting the beam by a field and so on. All of these operations support our belief that at least a large proportion, if not all, of the pulses recorded are due to electrons entering the counter. Since the electron is thus connected with the counter and we understand how the counter operates, the individual electron is regarded as existing at least in the event that is recorded. The place of the event is occupied by the detector, the time of the event is determined by using part of the electrical pulse operating the counter as the timing signal for an electronic chronometer. The events in which electrons are emitted from the source can be timed also, for the emission may be controlled by a pulse of voltage, or if the electron comes from a β-emitting nucleus, we could detect the pulse of ionisation produced by each nucleus as it recoils. The classical connexion between emission events and detection events can be established in a number of ways. For instance, we might measure the speed of the electron by passing it through a known magnetic field, limiting its path by slits. The timing of the detection pulse must be consistent with the time taken by the electron to reach the counter. This classical connexion conforms with what we should do when faced in practice with the need to devise a method to connect two appearances of a moving object which is imagined to traverse a track according to a particular kinematic law. The object existed in each event in which it was observed, and by hypothesis it existed in intervening events selected by the law.

Our discussion of interference and the uncertainty principle should have convinced us that this way of thinking about the connexion between the source event and the detection event is quite unacceptable when the electron undergoes diffraction. So far as physical observation is concerned, when interference is involved, there is no existence in the classical sense between the source event and that in which detection takes place. In the absence of a physical interaction by which the electron could be observed on the way, thus conforming to the requirements for interference, the source is connected with the detector in one jump. Observing the electron in between creates a different individual phenomenon.

Thus the motion that we are ready to treat crudely according to the classical method must be conceived as the transformation from one event to a later one. With the events, somehow or other an existence is connected. Classically, we want to put the thing in each event. The phenomenon of interference shows that we have to associate the existence with both events. The connexion between the events is an atomic existence.

Lest we are tempted to think that the electron so regarded is insubstantial, let us remember that the events we considered are real enough: they are supported by the apparatus we use to record them and by all the activities which as experimenters we have carried out to establish confidence in our observations, and also by the other activities which we could carry out to support our conclusions in the face of criticism or doubt. This condition of affairs is quite analogous to what goes on outside the physical laboratory in judging, and supporting judgment, to distinguish hallucinations or illusion or plain error from reality.

When we state that the connexion between the events is an atomic existence, we have in mind that it has the unity of the actual move of a piece on the chessboard. We could have placed the bishop at any of the intervening squares on the same diagonal row, but in doing this we should have made a different move. We do not think of the actual move taking place through a series of possible moves. Indeed, we usually lift the piece from the board and put it down on the previously unoccupied square that marks the new position of the piece. The elimination of the intervening possibilities on the board, or, to put the matter in a somewhat better way, the irrelevance of the intervening geometrical configurations through which the piece might be imagined to pass, is best illustrated by the knight's move. There is no knight's move in between. From a position in the centre of the board there are at most eight possible moves for the knight; these could be represented mathematically by an operator that transforms the pair of indices by which we number the squares of the board. Looking at the discrete system of possibilities of chess helps us to understand the implications of the word 'atom' in relation to existence and motion. In contrast to this, considering the electron being emitted and detected in a single jump does not conform with the classical view of the transition in terms of continued existence and continuous motion. In chess, we have no trouble seeing where the piece goes in making the move, for the space of possibilities for the game is only two-dimensional. In the 'move' of the electron, or other atomic particle, there is, however, no place for the electron to go. We cannot attribute to it continued existence between the source and the detector unless we invent a different space of more dimensions for its existence. Since this invention turns out indeed to involve a space of infinitely many dimensions in order to present motion as continuous, and since this has all the earmarks of mathematical fiction, we surely need to change our concept of existence. Whereas in classical representation we associate an existence with a single event, what we now require is to associate the existence with the pair of events connected by it. There is no motion between these events except in so far as we wish to call 'motion' the connexion between two events separate in space, and one (A) earlier than the other (B). If in the classical way we wish to interpolate an event (C) on the straight line connecting their spatial positions, we could have an existence AC, provided C represents a detection event properly related to the emission event A, but this would be a different atomic existence. It is not a part of AB any more than a possible move is a part of the existence of an actual move in chess. (We do not need to put the piece down!)

The events that are connected are in fact not point-instants, although we may for mathematical convenience represent them as such, subject, of course, to the limited precision allowed by the uncertainty principle in applying the representation. So long as the events are distinct and well separated in space and time, this limitation may be left out of account in discussing the existence of atomic entities, for our idea must depend on essentially classical supports as we have already discussed.

The concept we are discussing relates to the microphysical situations where classical ideas involve us in metaphysical difficulty concerning the experimental facts, but if we keep in mind the circumstances on which we depend in using classical ideas, we can see how the classical interpolation of events restores the ordinary conception of localised existence. Let us consider a set of classical interpolations C_1, C_2, ..., C_{n-1} between the events A and B. The connexions (AC_1), $(C_1 C_2)$, ..., $(C_{n-1}B)$ constitute a series of atomic existences. We therefore localise existences in this sense by enclosing the pair of events in each step in a small region of space, and think of the atomic existence located in it like an ordinary object, a mark corresponding to the middle of the region denoting the location. Since the classical representation of the motion of an electron or other microphysical object is successful only when the representation is sufficiently crude, it is seen that we can pass to the usage of ordinary language about the existence of the entity in space, provided that we support the usage in the ordinary way. An imagined interpolation, C, that is considered to have effect classically in actual fact, is incompatible with the existence that connects

A and B in one step. Thus we distinguish the *series* (α) of connexions (atomic existences)

$$(AC_1)(C_1 C_2) \ldots (C_{n-1} B)$$

from the *system* (β) of connexions

$$(AC_1)(AC_2) \ldots (AB).$$

The former (α) corresponds in the limit $n \to \infty$ to the classical concept of motion, the latter (β) does not. The system β requires us to invent the idea of something moving on different occasions, detectable only in a single event, in each case with a certain probability of detection. Motion is then represented by the particular classical kinematic law ordering the events in space and time. We are tempted to think we can still rely on the classical concept of physical existence, and because the object is an atom subject to quantum laws, speak as if the statistical effect of these laws permits us to think as we do of radiation.

The classical picture of radiation misleads us with a picture of something coming out of the source like a bullet from a gun. But to particularise the thing a label cannot be used. To name it one must point to the events connected by it. Here we are concerned only with the kinematic aspect of the matter. To establish what kind of radiation we are dealing with, we depend on our knowledge of the source of radiation and our understanding of the detector we use; otherwise we should not be justified in talking of light, X-rays, electrons, neutrons and so on.

Under the stimulus of our bullet analogy we think of things launched into space. In doing so we fail to appreciate that we are ordering independent atomic existences into a series with a common initiating event (or one imagined to be so) and a variable terminating event. To get to a particular termination, however, we do not go through the intervening members of the series, unless there are physical effects actually occurring by which to label the track. This, of course, bears a resemblance to classical observation of a moving object.

A particle of a radiation field connects an emission event with a detection event; it may, of course, also involve other events. The existence at the emission event is the recoil, or some such physical effect in the emitter, that can be observed. The recoil really does take place, whether we actually observe or not: that is, the system of physics requires us to think of the matter in this way. In spite of the attempts by some philosophical writers to make it appear that unless the effect is observed we cannot say that it has taken place, no physicist is going to doubt the recoil of a β-emitting nucleus whether he has made the experimental arrangement to observe it or not. Indeed, if he failed to observe the recoil, he would start looking for another participant in the process to supply the amount of momentum required to balance accounts. Just as the emission event is connected with an existence, so is the detection event, for example the pulse of ionisation in a gas counter. These two events, as it were, bound the existence of a particle which is imagined to leave the source and enter the detector and is thought of as moving in the intervening space. Provided that the representation is crude enough this view is practically effective.

If sharper representation is attempted, the classical view of motion and the attendant possibilities of interpolating observations that do not disturb the physical process of the motion being observed are no longer allowed. Unless we accept the transition between the events at the source and detector as the existence of the entity concerned, we are not on safe ground in relation to the interference of probability. In so far as we can ignore the physical effects due to the presence of distant objects, we can represent the entity as localised in an extended region that includes both events— but then we cannot have it moving within the region in any classically acceptable sense. Only if we can interpose events at which physical interaction takes place, that can be observed, is the picture of classical motion applicable. Thus motion in quantum mechanics with non-classical statistics is represented by a set of separate existences which overlap on our map, whereas what corresponds to the classical case is a series of consecutive jumps in a continuing existence. The distinction between these two was made in discussing the motion of a spot of light along a row of lamps.*

* Chapter V.

94

In quantum mechanics the variable termination becomes the position of the moving atomic particle in the continuum. It should be obvious that if we wish to represent the variable termination as a moving point governed by a continuous law of motion, we cannot have the point move in ordinary physical space, because we must eliminate the intervening possibilities of ordinary physical space between the source and the detector.

In order to accomplish this, we have to treat in a new way the points of space which serve as possible terminations of an atomic existence in the system we are considering. We must treat them as independent logical possibilities which, while corresponding one-one with the possibilities of physical space, are not presented in a system that imposes spatial order in the way kinematic order applies to the point-instants in ordinary space-time. It must be possible to pass in the new space from event A to event B without necessarily passing through the event C which we would think of physically as lying on the possible path of a particle through A and B. We therefore treat the system of logical possibilities as we should treat the points of a lattice in relating the lattice to the continuum, for there we meet exactly the problem we are considering. On the lattice, there is no possibility between adjacent points. When the lattice is presented in the continuum, however, points of the continuum, lying between those corresponding to adjacent points of the lattice, do not correspond to points of the lattice in the ordinary way. It is possible to join any two points of the continuum, that correspond to lattice points by means of a continuous line that passes through no other points of the continuum corresponding to points of the lattice. But this way of thinking about the matter will not really serve because it connects the points of the lattice through the continuum. We may exhibit in an elementary way what is required.

Consider the intersection of a rectangular axis triad ($OABC$) with a sphere of unit radius centred at O. Let the points of intersection be a, a', b, b', c, c'. The directions $Oa, Oa', Ob, Ob', Oc, Oc'$ stand for six distinct logical possibilities with which we are all familiar in ordinary language, viz. up–down, right–left, fore and aft.

95

Vectors joining O to points (x) of the sphere other than a, a', b, b', c, c' do not stand for logical possibilities in this system. The vector Ox is a linear combination of the basis vectors Oa, Ob, Oc;* it does not represent one of the six logical possibilities, but it could represent a statistical combination of them, provided that we have regard to the laws of probability in establishing the representation. The mathematical elaboration of this is obvious since the sum of the squares of the components of Ox is unity. The important point for our purpose here is that some points x are traversed by the representative point on the sphere in continuous transitions between pairs of the six logical elements of our system. A jump in the discrete system is then regarded as a continuous change according to some law which we have not yet specified. Once the continuous law is given, however, a track is specified on the sphere, and in going to Ob from Oa, for instance, would pass continuously through points between Oa and Ob on the track.

This scheme of representation may be extended to a system of n possibilities and a representation in n-dimensional linear vector space. Imagine possibilities for an existence on a lattice (h, k, l), where h, k, and l are integers. We shall denote a variable point of the lattice by the integer triad (ξ, η, ζ). The mathematical identity

$$u(\xi, \eta, \zeta) = \sum_{h, k, l} u(h, k, l)\, \delta_{\xi h} \delta_{\eta k} \delta_{\zeta l},$$

$0 < h, k, l \leqslant N$, where δ_{mn} is the Kronecker delta symbol, can be read as a vector equation in the space in which each

$$e_{hkl} = \delta_{\xi h} \delta_{\eta k} \delta_{\zeta l}$$

is a basis unit vector with $e_{hkl} \cdot e_{h'k'l'} = 0$ or 1 according as the corresponding primed and unprimed subscripts differ or agree. $u(h, k, l)$ is the hkl-component of the vector $u(\xi, \eta, \zeta)$; it may be a continuously variable number of absolute magnitude not greater than 1, and indeed, corresponding to our previous spherical representation

$$|u(\xi, \eta, \zeta)|^2 = \sum_{h, k, l} |u(h, k, l)|^2 = 1.$$

* $Oa + Oa' = 0$, where $+$ stands for the logical 'or'. So $Oa' = -Oa$. This is a necessary part of the logic of the system since Oa and Oa' are mutually exclusive.

Thus we have a N^3-dimensional linear vector space in which $u(\xi, \eta, \zeta)$ is represented as a vector.

Now consider displacement from (o, o, o) to (a, b, c) *on the lattice*, that is, through points (ξ, η, ζ) for which the succession of coordinates is governed by the following rule (A): in each step only one of the integers ξ, η, ζ changes by unity, while the other two remain unchanged. On the lattice this resembles continuous motion in a continuum.

The corresponding transformation from $u(o, o, o)$ to $u(a, b, c)$ in the vector space proceeds via rotation from one coordinate axis to the next. But clearly if u_{hkl} and $u_{h+1, kl}$ are positive numbers

$$u(\xi, \eta, \zeta) = u_{hkl}\, \delta_{\xi h}\delta_{\eta k}\delta_{\zeta l} + u_{h+1,\, kl}\, \delta_{\xi h+1}\delta_{\eta k}\delta_{\zeta l}$$

represents a vector *between* e_{hkl} and $e_{h+1,\, kl}$. That is, whereas on the lattice we are concerned with integral steps between points of the lattice, in the vector space we can prescribe a continuous transition through the corresponding basis vectors in turn. Further, if we wish to allow only the integral connexion (o, o, o) $\rightarrow$ (a, b, c) on the lattice, we may prescribe a continuous one-parameter family of vectors bounded by $u(o, o, o)$ and $u(a, b, c)$ which includes none of the basis vectors except e_{000} and e_{abc}.

Thus this representation makes possible continuous transition (or motion) in the vector space corresponding to the connexion between two widely separated points of the lattice without requiring that intervening points of the lattice be traversed in accordance with the rule A.

Let us now pass to the representation of possibilities for existence in the continuum of physical space by letting N tend to infinity. The elementary approach to this idea depends on regarding the argument (x) of a function $f(x)$ as an index. Each index denotes a logical possibility which can be represented as the unit displacement along one of the axes of a coordinate system. Each value of $f(x)$ is then regarded as the component of a vector. The vector represents $f(x)$ in function space, the component is taken along the axis corresponding to the particular value of x, so $f(x)$ becomes a logical symbol which is expressed as a linear combination of the basis vectors. The following table shows the analogy.

	3-dimensions	n-dimensions	$n \to \infty$
Basis vectors	$\mathbf{i}, \mathbf{j}, \mathbf{k}$	e_1, e_2, e_n	$\delta(x-\xi)$
Components	x, y, z	x_1, x_2, x_n	$f(\xi)$
Vectors	$\mathbf{r} = x\mathbf{i}+y\mathbf{j}+z\mathbf{k}$	$\mathbf{r} = \sum\limits_{s=1}^{n} x_s e_s$	$f(x) = \int_{-\infty}^{\infty} f(\xi)\delta(x-\xi)\,d\xi$

In the infinite dimensional column ξ plays the role of the subscript, or index, s, of the column for n-dimensions. The basis vector is a logical symbol, the general vector being a sum of components is a logical combination.

The basis vectors for representing the possibility of existence in the continuum of physical space are

$$e_{xyz} = \delta(\xi-x)\,\delta(\eta-y)\,\delta(\zeta-z);$$

each spatial position (x, y, z) names one logical possibility. The distribution $f(\xi, \eta, \zeta)$ is represented as a vector whose component corresponding to the basis vector e_{xyz} is $f(x, y, z)$; in place of the sum in the finite-dimensional vector space we now have the integral

$$f(\xi, \eta, \zeta) = \iiint f(x, y, z)\,\delta(\xi-x)\,\delta(\eta-y)\,\delta(\zeta-z)\,dx\,dy\,dz.$$

The integral connexion between two points of physical space, one corresponding to the source and the other to the detector of a physical particle, is represented in the vector space by a continuous transition which need not require the vector to pass through any of the basis vectors.

In quantum mechanics, the domain of numbers in which the components of the vectors vary is not the continuum of real numbers hitherto assumed above, but that of complex numbers. Mathematical convenience was offered by Weyl as sufficient grounds for this,* but it is more reasonable that we should connect this choice with physics and our concept of atomicity.

In representing the emission of an atomic particle from a source we should normally choose as the initial vector, consistent with our knowledge of the emitter, one of the characteristic outgoing

* H. Weyl, *The Theory of Groups and Quantum Mechanics*, trans. by H. P. Robertson, ch. 1, §4 (New York: Dover Publications, Inc.).

spherical waves (an eigenfunction of the angular momentum). The location of the source is then indefinite to the extent determined by a momentum uncertainty equal to twice the momentum of the emitted particle. But when the source is observed to recoil we have to choose a different vector to represent the outgoing particle: because of our enforced ignorance in the circumstances we have imagined, it cannot be an eigenvector of position or momentum, and it cannot be one of angular momentum. The particular vector judged appropriate for our representation carries the information needed when we apply the dynamical law to determine the probability of the free particle being detected by a suitable detector placed in space to receive it. Thus the unitary vector associates a probability of detection with each point of physical space representing the position of a possible detector, and consequently with each member of the *family* of atomic existences that can be associated with the event in which the source recoils in a particular direction and 'a particle is emitted'.

If some diffracting or other physical system is interposed between source and detector, the 'state vector' for the particle is transformed by dynamical interaction and the new probability for detection at a given place is determined by the modulus squared of the component of the transformed vector in the functional direction $\delta(x - x_D)\, \delta(y - y_D)\, \delta(z - z_D)$, (x_D, y_D, z_D) being the position of the detector.

In the functional space each possibility of existence is represented by a ray of 'unit length' from the origin. Only the particular axes that are eigenfunctions of 'the particle's positional coordinates' correspond to the existence in physical space. Any other ray does not correspond to a point of physical space at all because it involves in a sense *every* point of physical space. Thus in the functional space we have possibilities for existence that are not possibilities for existence in ordinary space. They can be interpreted statistically, however, and this is what quantum mechanics does.

In classical representation the physical thing is located in the event. This act in the representing process may stand for a quite different and more elaborate pattern of behaviour than merely marking the map, for it involves actual things and other processes. Whatever is done must be acceptable as a support for attaching to the event a name by which an existence inheres in the event. According to the conception presented here for atomic existence we should attach the name to the atomic transition between source event and detection event. In a crude representation in which these events are not distinguishable the name is in effect attached to a single event, as required by the classical representation with which we use the word 'existence' in ordinary language.

We have spoken of the connexion of events, but the particles of modern physics establish other connexions. For example, consider the diffraction of a molecular beam. The latter is limited by a nozzle and a slit system, and the crystal which diffracts the beam has a finite size. The detector may be a chemical detector which shows the capture of molecules either by change of colour or through development, or it may be a very fine tube which admits molecules to a reservoir in which pressure change is measured. Here the atomic connexion is between the incident beam classically conceived and the diffracted beam defined by the detector also conceived classically. The connexion now is between values of the classical momentum which, of course, are not specified with limiting precision, any more than the events of which we have already spoken.

We examine the experimental arrangement and the processes used by the physicist in order to see how the atomic existence is involved. In our connecting parts of our representation of which we have classical images, which we approach metaphysically through ordinary language, and whose existence is verifiable in the ordinary ways, the atomic existence is itself established as an existence that connects the images in an atomic step. Its existence is supported by the existences it connects. As an atomic step it does not encounter the question 'what happens in between?'. But what do we mean by 'connecting'? Surely it is not only a connexion in thought: there must be some physical change or transformation involved that we may measure or infer from measurement.

Electric charge, mass, energy and momentum may be transferred. In the molecular beam experiment we have to deal with the transfer of gas from the source of the molecules to the high vacuum detector. When neutrons are diffracted, the beam may be derived from the thermal column of a nuclear reactor, and they are detected individually by the pulses of ionisation resulting from neutron capture in nuclei of boron in the gas filling the ionisation chamber. Neutrons and molecules of well-defined speed may be oolootod by a timo of flight opootromotor whioh io oooontially classical in conception. And so on.

Now the association of an emission event with a capture or detection event is not a simple matter in practice. Provided that not too many events are observed concurrently in the same region of space, it is usually possible by plausible classical reasoning to connect the events in a physically acceptable way. But not always. The processes on which we depend work by bracketing in space and time, if possible so as to select a single detection event that is consistent with what has been measured at the emission event. For instance, good information about the direction of recoil of the source would limit the locus of possible subsequent interaction to a certain solid angle, and if the initial momentum is known well, the range at which interaction could take place is selected by timing, or by observing a process from which timing could be inferred.

It is when we do not have these aids, or when we think about atomic processes under the strong influence of the idea that the waves represent an existence, that we have great difficulty in accepting the idea of any connexion at all between the emission event and the detection event. The emission event then contributes to the field and the detection event appears as an unrelated spontaneous action by the field on the detector. The logical situation resembles dropping a coloured ball into a bag in which there are already present numbers of balls of the same size, but of various colours. A ball is withdrawn from the bag: it has the same size and colour as that dropped into it, but there is no mark or continued observation through which we might discriminate whether the ball that is taken out is the same as that put in. That is, in the physical situation we must look at the circumstances of the case to see if we have reasonable grounds for speaking of the particle emitted from A having been detected at B. We must be able to establish the connexion in some way other than by merely asserting it. It is for this reason that an interpretation of events observed in photographs of tracks in a cloud chamber, an emulsion, or bubble chamber must have physical effect. The specks along the track are seen as connected by a moving particle. Nuclear events and elementary particle disintegrations are seen as connected, because, dynamically, the physical quantities that are measured fit the interpretation, even though we have to imagine that the gaps in the picture are crossed by neutral particles that leave no track, or by charged particles that pass without causing any ionisation to form drops or bubbles. These practices surely show that the claims by theorist about only treating what can be observed or measured are not consistent with the plausible reasoning of experimental scientists.

Thus we see that using the ideas of modern theory in treating experimental observations may be much more subtle than can be formally organised in a set of axioms or principles about observing nature. The experimenter does what he can under the circumstances, offering hypotheses as the expression of the interpretations he is disposed to favour. The determination that these hypotheses are acceptable to others takes time and a variety of activity and experience by other scientists. Winning assent is a much more involved process than merely presenting a logical argument and exhibiting the compatibility of the theoretical estimates with measured magnitudes. The formal exposition of theory in books is directed to exploring for the learner the forms of thinking that have already won assent, and tends to conceal the real processes which guided men to commit themselves to the particular inventions expounded in them. The more elaborate a theory becomes, the more difficult is it to explain the grounds for our confidence in it. Especially is this the case in the face of philosophical questions that can be properly answered only by taking into account how the theory is used in contexts where ordinary language, and our habits of thinking in using it, are taken for granted, or are displaced in the theory by formal substitutes that seem essential to support the forms of expression and the mathematical system of the theory.

In quantum field theory we find explicit expressions combining, for example, the creation of an electron say at x_1, its propagation to x_2 and its annihilation in the event x_2, which must lie in the absolute future of x_1. Neither of the imagined events may be observable, yet the theory connects them for the purpose of making the statistical calculation in which x_1 and x_2 are variable. Further, this particular connexion may be absent from alternative evolutions of the various wave functions of the system that must be combined coherently. The electron or other elementary particle, if it exists at all, exists as a connexion between variable events none of which need be observed. Yet elementary particles of the type in question have been undoubtedly observed to connect events in other experiments by the same laws assumed in field theory. Accordingly this kind of existence is far removed from the simple types of existence we have already discussed, where the events connected by the atomic existence are directly supported by the real world of ordinary experience. The supports for the imagined transitory existence have to be extended into the past of the initiating event, and into the future of the terminating event, until we encounter the substance of ordinary matter. There is nothing feeble about this extension of support, but we must note the philosophical point that the case is different from the simple one, and the justification of the former differs from that of the latter. Logically what is going on is quite analogous to the variety of method in surveying and in measuring inaccessible objects in science. We invent a fairly elaborate procedure for reaching what we want, making use of theory to extend our practical knowledge; and we have methods for judging the precision and reliability of procedures. In the light of such reasonable operations, presenting physical theory without admitting the possibility of them seems to put us in an intellectual strait-jacket.

We have already discussed on several occasions the idea that an atomic existence connects events. If the events are close together we can, in order to get a classical view of the existence in space, think of the particle as occupying an event in a region that includes the pair of events and is crudely specified. The existence of the entity is to be conceived in space-time with this condition. Our real difficulty arises when we think of the precisely mapped environment in which the connexion between the events is set up. Here the precision is relevant to the wave calculation. But this is not pertinent to existence. It has to do with probability applied to events at different places in space. The propagation of the waves is determined in the calculation by physical properties of the environment. These properties therefore determine for the elementary particle only a statistic of possible existence. By means of the waves they cannot determine anything else, for the waves represent the various possibilities, the independent channels, over which the theory establishes the probability distribution. If we wish the waves to represent the atomic particle leaving its source, being diffracted, and then entering a detector, we must assume a particular characteristic wave emanating from the source, and a particular characteristic wave at the detector, each coupled to the system of waves duly diffracted in between. This is necessary in order to achieve the representation by means of a particle going from the source and caught by the detector and transferring mechanical and other quantities from source to detector.

In a somewhat different way, however, we have to think of standing waves of light in a maser. The atoms which supply energy to the light field by stimulated emission are coupled to the various optical modes. The strength of this coupling depends on the position of the atom in the 'resonating cavity'. It determines the probability per unit time that the atom will be de-excited and transfer its energy to the appropriate standing wave mode. No more detailed picture of the photon transfer is physically effective.

Let us return to our intuition that there *must* be a process intervening between events connected by an atomic existence. We are ready to accept the idea that this is the wave process of wave mechanics because formally it enables us to manage microphysical phenomena. But in using it we have to keep reminding ourselves that the waves do not transport energy and momentum in the way that a classical wave motion does. The quantisation of energy, for instance, forces us to interpret the continuous physical magnitudes of the field with the aid of probability. That is, even if we introduce the wave process as the physical connexion between the events we are no nearer achieving our goal. In the tradition of aether theories we are looking for substance and existence in the wrong place. Our continuous space-time map is much too convenient for mathematical representation. We forget just how elaborate are the processes by which we gather the information represented in the map. An elementary particle is placed in an environment to describe which we have to depend on resources not available to the particle. We present it with a much more complicated world than it has the means of resolving in detail—think of the limitation on what we ourselves see in the field of a microscope due to the finite wavelength of the light. Thus we lose the one-to-one correspondence between the structure of the environment explored by the manifold resources of the experimenter (and the technology he depends on) and the structure of the response of the atomic entity, except through the statistical correspondence of modern theory. We are forced to this kind of correspondence because we are dealing with atoms. But this thought does not touch, except very superficially, the understanding of how Planck's constant dominates the way in which the atomic entity looks at the possibilities of our continuous space-time map. We provide the entity with infinite possibility in our continuous representation which we then have to restrict by the quantum laws, and the special logical categories of which they treat. For example, the angular momentum of our atom in an experiment to measure it may be found to take one of a discrete series of values. But there is no necessity that under other circumstances we must represent the atom's angular momentum as determined in the foregoing way. We admit the possibility of a continuously variable expectation value of the angular momentum and interpret it as the statistical average over the discrete series.

An atomic existence involves space as a single channel. If we attempt to resolve it into subchannels we destroy it and create different existences. With the aid of experimental details about the emitting and receiving events, the channel may be much better defined in regard to its dynamical aspects than it would be without them. For instance, the detection event may involve a particular nuclear or atomic transition which we know requires a particular angular momentum change. The incoming wave therefore is restricted to an angular distribution of amplitude and phase that conforms with this change. Again, the receiving atom or nucleus may be oriented in space. This imposes further restriction on the incoming wave. Thus, as it were, by our choice of method for observing nature we impose form on the phenomena we observe. We pick out particular forms. If our arrangement for detection is not discriminating enough, we have to deal with an expectation value which is a (quantum) statistical average. Thus waves entering the calculation in this case are less well defined than in more discriminating circumstances. But surely we are not going to say on that account that there exist two (or more) wave systems which are different according to our knowledge of the properties of the detecting system. Yet this kind of statement has been made: as if there were no physical process going on without our intervention. The essential property of the waves is that, being statistical in effect, they cope with variety of instance and are therefore in some important respects independent of the particular instance of detection. We may think of the detector acting on the waves in its particular way, but if we are to think of using our detector for the purpose of studying a physical phenomenon, we must limit the extent to which the detector participates in it so that we can be 'sure' that its role in what we observe is understood and does not perturb unnecessarily the process being investigated. In this matter we cannot draw sharp boundaries about procedure: and the grounds on which one physicist may be confident in his interpretation may differ from those of another, and indeed both may be reinforced with decisive effect by the observations of others on a different phenomenon.

What have the waves to do with a single atomic existence? Their role concerns the statistical structure of families of related existences that we want to interpret by means of the idea of motion. Since the waves have this role they deal essentially with inter-relations of possible existences in the systems prescribed by our physical models in the modern theory. They are thus external to the existences. This is a conclusion that is in striking contrast to the idea we have reflected on several times, namely that the waves bring into the representation the statistical effect of what stands between the initial event and the final event as if we were representing the diffracting system with a finer resolution than is permitted by the uncertainty principle.

Thus what has the appearance of representing a process between the events initiating and terminating an atomic existence really stands for the organisation of a family (or families) of possible atomic existences. That one member of the family terminates in an event earlier than that terminating another member does not put the former physically inside the latter. Motion is a property of the family. This is all very well for the unbound particle. What about an electron in an atom, for example? In a stationary state the waves are standing waves when the 'orbital' angular momentum is zero. If not, the classical concept of a revolving charge is mirrored in the eigenfunction's complex dependence on the longitude ϕ. The atomic event in this context is an atomic transition accompanying the absorption, scattering or emission of radiation, or interaction between the atom and some other piece of matter. The existence of the atom is not transitory in this context. It can support an interaction with radiation or excitation due to collision, and it may enter into chemical processes that alter it in various ways. These, however, do not put its existence and endurance in question in the contexts mentioned.

Nevertheless, we have to face the question of motion of atoms in the dynamical representation of the properties of matter. Surely we require a concept of the existence of the atoms of helium in the liquid at low temperature. So far as the liquid in bulk is concerned, in spite of its remarkable properties, it looks, and in many respects we can treat it physically, like an ordinary liquid. Yet we know that atomic beams of helium can be diffracted. Are we therefore to speak of the motion of an atom in such a beam in the same way we use for radiation? Are we to say that there is no 'between' for the atom as we did for the electron? If we wish to exhibit the phenomenon of interference, the answer is 'yes'.

Let us think of the Estermann and Stern experiment—the diffraction of hydrogen molecules by a crystal. The diffraction phenomenon involves not merely incident and emergent waves from the surface grating but also the system of evanescent waves close to the crystal grating. These represent states of motion of the molecules under the action of the forces between the molecule and the atoms of the grating. The three systems are coupled together—incident, evanescent and emergent, and some of the evanescent modes represent states in which the molecule is held on the crystal surface. So the presence of the crystal is represented in the structure of the wave field in space. The molecules could be detected there, but this is merely a statistical alternative to molecules entering the diffraction pattern.

The molecules are incident on the small crystal in a limited beam, and the diffraction is observed in an enclosed evacuated space. We have no doubt about the existence of the hydrogen molecules in this apparatus, or that they leave the source of the beam and are caught in the detector, or on the surface of solid objects in the vacuum, or removed by the pump. These ideas conform to classical conceptions. But the connexion between the beam and the distribution of diffracted intensity does not, and to explain this connexion no classical limitation in our thinking is appropriate. The properties that we wish to attach to the individual molecule in thinking about it are associated in a very complex way with the physical arrangement to produce the beam, and the physical effects by which the molecules are detected. These physical arrangements comprise the source of hydrogen gas, control of the temperature and pressure of the gas in the reservoir connected to the nozzle by which the molecular beam is produced and directed on the diffracting crystal, the detecting system which captures the diffracted molecules and by the particular circumstances of its use, records not merely the rate of capture of molecules, but also establishes that they are hydrogen molecules received when the beam is incident on the crystal. The beam is conceived as a classical agent and so is the sample of diffracted molecules—it affects the detector. But the classical relation in which these two stand in space is not applicable to the individual molecules for this relation is a statistical one.

The properties we have been discussing are attached to the atomic connexions between the beams, not to a particle of ordinary matter travelling through space. The continued existence of the molecule in diffraction is not to be thought of in the sense of classical motion. Its existence is an atomic connexion between the beam conceived classically and the detector as a classical device, and since it is an atomic connexion the idea of continuous existence in intervening space and time, although pertinent to classical representation, is excluded.

We can imagine the diffraction pattern to be recorded by means of a chemical detector so that the molecules in the diffraction pattern are laid down on a plate and endure there for development. We have no doubt of their existence there. Why, then, we may ask, do we seem to have the molecule somehow or other disembodied in its motion? Surely the answer is that the continued existence of the molecule on the plate is conceived in a classical context without the aid of the resolving power of the atomic grating. While our classical conception is applicable to the crude description of the motion of molecules or atoms, it is not applicable to the atomic steps governed by diffraction, for under the conditions for interference we cannot interpolate in the classical way.

CHAPTER VII

THE CONTINUUM AND
THE LATTICE

Physical space is conceived as continuous and every other space we wish to adjoin to it for representing physical processes by classical methods is likewise regarded as a continuum. The coordinates by which we name possibilities in these spaces are continuous variables; we assert that we can go on indefinitely interpolating names with greater and greater refinement. Since the invention of the differential calculus, mathematical analysis has evolved to provide powerful tools that aid thinking about physical representation by means of continua. Analytical devices have been invented so that not merely continuous distributions in space and time can be managed by continuous analysis, but also the discrete and discontinuous can be treated with formal facility by means of complex variables.

Quantum phenomena exhibit spectra of discrete values for the variables that on the basis of classical physics we wish to represent in a continuum. The atom of action, introduced by Planck when he explained the distribution with respect to wavelength of the energy radiated from a thermal enclosure, stands for the least action interval, or step, permitted between distinct values of the physical variable, action. The essence of atomicity is that a limit is set to interpolation—for we cannot divide an atom; it cannot be cut. But setting a limit to interpolation may enter physical theory in quite subtle ways. We must find the right variables to which it applies. For instance, for the stationary states of an atom one such variable is the energy. But the spectrum of allowed values of the energy is not evenly spaced in energy-space: it is the quantum numbers that exhibit uniformly equal steps between integral or half-integral values.

In representing the discrete values nh for the action, where n is an integer, our classical habit disposes us to mark a series of equidistant points on a continuous line. Indeed, when we state that a numerical magnitude can assume only integral values, we are already placing it in a system that presents the possibility of non-integral values. We rely on the symbol that stands for the variable magnitude, and are ready to use it in algebraic operations that define an algebraic field. The number of particles counted in a physical observation is such a magnitude. Non-integral values stand for statistical averages because, in computing them, we could generate the fractional part by arithmetical division.

The foregoing comment seems to indicate how we may interpret the intervening points of the continuum in the continuous line joining the discrete series of points. By extending the process to a number of dimensions we can grasp the possible role of statistics in connecting representation on a continuum with the corresponding lattice representation. It is an important role because we are led by mathematical convenience to depend on the continuum in physics. Since we insist on using continuous representation, we must introduce a statistical or other interpretation to relate the continuous possibility to the restricted possibility of the lattice. A point in the continuum lying between lattice points may be presented by means of a linear combination of vectors, each of which corresponds uniquely through its direction to a possibility on the lattice, as we discussed in connexion with atomic existence.

Let us now consider a different way of relating the continuum to the lattice. Suppose we are dealing with a variable R that may take the values $0, 1, 2, \ldots$ We mark these on a line at equal intervals so presenting the series of allowed values joined by the intervening continuum of the line. Now let us mark the points of a second set $\alpha, 1+\alpha, 2+\alpha, \ldots$, where α is a positive fraction less than unity. This set of points is derived by displacing the former set through α in the positive sense. As a representation of the set of integral values, the second set is equivalent to the first in that the step between successive points is unity. In order to exhibit the restricted possibility of the discrete set of points, and yet attach significance to the intermediate points of the continuum, we shall represent each point of the discrete series by an axis in function space. In the system we are considering, the basis vectors of the first set are $\delta(R-n)$, where n is an integer, and for the second $\delta(R-n-\alpha)$. These form two distinct subspaces of the function space required to present the possibilities of the continuum. As α is varied from o to 1, however, the basis vectors of the second set are 'rotated' in function space, and when $\alpha = 1$, coincide with the set corresponding to $\alpha = 0$. In elementary terms, the complex number of unit modulus $z = e^{i\theta}$ is represented on the xy-plane through $z = x+iy$. As θ is changed, the vector representing z is rotated; but when θ changes by 2π, then z is represented by the same vector (unless we think of representation on the Riemann surface of which the successive sheets distinguish complete rotations about the origin of z). What we have to deal with is the generalisation of this idea to function space. Corresponding to the rotation on the complex plane we have unitary transformations of function space.

We shall distinguish the equivalent series of points by means of the subscript α; thus the points of R_α correspond to the basis vectors $\delta(s-n-\alpha)$ of the continuum (s). Because of the equivalence of the sets R_α by which we may represent the discrete series for R, we have to adapt our representation. This we do by resorting to the domain of complex numbers.

The reader is reminded that in the preceding chapter it was indicated that a function $f(s)$ may be regarded as a distribution over the singular possibilities of the variable s; thus

$$f(s) = \int_{-\infty}^{\infty} f(\xi) \, \delta(s-\xi) \, d\xi$$

presents $f(s)$ as the logical combination of the possibilities $\delta(s-\xi)$ with amplitude $f(\xi)$. But it could also be presented as a distribution over a complete system of independent possibilities $\phi_\sigma(s)$; these are orthogonal normalised functions of s and

$$f(s) = \sum_\sigma c_\sigma \phi_\sigma(s),$$

where $\sum_\sigma$ stands for summation over σ, and may involve integration over a continuum of σ as well as summation over a discrete series. This form for $f(s)$ corresponds to a new choice of basis vectors, namely, $\phi_\sigma(s)$ in place of $\delta(s-\xi)$. Thus the lattice of s values restricted to integers can be related to other systems of logical possibilities that involve the entire range of s through distributions with characteristic forms.

Let us transform to a different system of basis vectors by means of Fourier transformation. For this purpose let σ be the variable conjugate to s. The new basis vectors are the functions $e^{i\sigma s}$, one corresponding to each value of σ. Since

$$\delta(s-n) = \frac{1}{2\pi} \int_{-\infty}^{\infty} e^{i\sigma(s-n)} d\sigma,$$

the basis vector $\delta(s-n)$ has the component $(1/2\pi)\,e^{-i\sigma n}$ with respect to the axis $e^{i\sigma s}$ corresponding to σ. Similarly, $\delta(s-n-\alpha)$ has components $(1/2\pi)\,e^{-i\sigma(n+\alpha)}$ .Thus the distribution $e^{i\sigma\alpha}\,\delta(s-n-\alpha)$ over s and σ has components $(1/2\pi)\,e^{-i\sigma n}$ which are the components of $\delta(s-n)$, and so $e^{i\sigma\alpha}\delta(s-n-\alpha)$ represents the same logical possibility as $\delta(s-n)$. Correspondingly $e^{-i\sigma\alpha}\,\delta(s-n)$ represents the same logical possibility as $\delta(s-n-\alpha)$.* Thus associating the phase $\sigma\alpha$ with the lattice R_0 in a complex number enables us to pass to the equivalent discrete set of points R_α.†

The two lattices R_0 and R_α correspond respectively to different subspaces of function space. One may pass from one to the other by a unitary transformation which may be regarded as the formal generalisation of rotation in the plane on which we represent complex numbers.

In order to exhibit in principle what is involved, let us think of the 12-hour clock face graduated in minutes, and let us look at the hour hand only. Imagine that we have observed the frequency of some particular happening on the hour, so that we attach to each hour mark a number that represents the probability of the happening. By means of the hand-setting mechanism, we could displace the incidence of the observed occurrences to the times

---

* From the point of view of operator calculus, since $i\sigma$ can be represented as d/ds, this result is formally equivalent to Taylor's theorem.

† Cf. the statement that altering the phase of the state vector in quantum mechanics does not change the physical situation represented by that vector.

denoted on the clock face by $\frac{1}{5}$, $\frac{2}{5}$, $\frac{3}{5}$, and $\frac{4}{5}$ of an hour from the former indications. Thus we place the set of 12 hourly possibilities in a system of 60 possibilities. There are five such distinct sets in the system of 60. On the clock face, rotation through the angles 6°, 12°, 18°, or 24° transforms one set into each of the others in turn. In the function space, however, each set of possibilities requires a subspace of 12 dimensions in the system of 60 dimensions.

The probability distribution which we introduced above is represented by a vector of unit length in the subspace corresponding to the indications on the clock at the exact hour; the direction of this vector is given by its components, each of which is the square root of the probability associated with the corresponding hour. In order to pass to a representation with a different setting of the clock, we use the appropriate one of the transformations of the 60-dimensional space which transforms one subspace into another, without altering the lengths of the vectors. The transformation required is a unitary transformation, and its inverse is also a unitary transformation.

By superposing the distributions which are obtained by applying the four* transformations individually, and normalising so that the resultant vector in 60-dimensional space has unit length, we generate a distribution in that space that attaches in general a complex number to each indication of the clock face. But the structure of this distribution could be analysed to invert the process and reveal that in fact we have to deal with happenings each hour.

For the purpose of achieving a concrete grasp of what is involved in these considerations, placing one finite system in another larger finite system is helpful. Its explicit mathematical expression, however, appears much more elaborate than the corresponding expression by which we implant the lattice in the continuum, because in the latter, we can depend on Fourier transformation; but this is merely a matter of simplicity of notation, not of intrinsic simplicity in conception.

We noted above that $e^{-i\sigma\alpha}\delta(s-n)$ represents the same logical

* Five, if we include identity.

114

possibility as $\delta(s-n-\alpha)$, and $e^{i\sigma\alpha}\,\delta(s-n-\alpha)$ the same possibility as $\delta(s-n)$. These expressions exhibit what has been indicated about the effect of unitary transformations. They require that, in giving effect to them, we must know what matrix* form to give to σ in order to produce the appropriate unitary transformation $e^{i\sigma\alpha}$, and its inverse $e^{-i\sigma\alpha}$, which transform function space, turning one lattice into the other. Thus relating the lattice to the continuum introduces an element which we did not expect when we set out on this discussion. We have revealed that complex numbers must come into the picture in order to represent the equivalence of the lattices as systems of possibilities, and we have also revealed that the variable σ conjugate to s with respect to Fourier transformation plays an essential role in determining the unitary transformations. In fact, σ must enter because we are concerned logically not merely with distributions on s, but also with displacing them on s.

When α is made infinitesimal $(=ds)$ the matrix $e^{i\sigma\,ds} \to 1 + i\sigma\,ds$. The change in a vector X which represents a distribution over s is

$$X' - X = i\,ds\,\sigma X \quad \text{or} \quad \frac{dX}{ds} = i\sigma X.$$

This is a vector equation that shows the infinitesimal transformation of the vector X corresponding to the displacement of the lattice by ds.

So we cannot specify the matrix of the transformation to which σ corresponds until we know how to transform the distribution X as we displace the lattice. In this way we recognise that merely to assert that a variable takes a discrete set of values which we set down in the corresponding continuum is only part of the representation of atomicity with respect to that variable. To use the continuous representation, we must introduce also an idea to guide us

* For a unitary transformation, σ must be a Hermitian matrix. A unitary transformation does not alter the eigenvalues of an operator represented by a Hermitian form. So in fact the continuous system of possibilities may be regarded as generated by unitary transformation of the discrete system. In going to the continuum we do not introduce new possibilities for the variable which we regarded as completely specified; we fill the continuous space with equivalent representations. In this process we depend on complex numbers to achieve sets of subspaces connected by unitary transformation, each of which serves to represent the logical possibilities of the original system.

in choosing systems that determine appropriate forms for the distributions we place at the intervening points of the continuum.

Our discussion shows us also that the variable action (S) which Planck quantised must be associated with a conjugate variable in order to give the law for transforming the distribution over the integral values of S to intervening points of the continuum. The physical significance of this second variable is exposed in the quantum theory of boson fields.* Associated with the phase ϕ_λ of the wave function representing the particle state λ, defined by the components of the momentum of the particles, is the number of particles N_λ. These two variables ϕ_λ and N_λ are connected with each other like our s and σ, and ϕ_λ is in fact a difference in action divided by $\hbar$. The variable N_λ corresponds to the gradient of the distribution over the phase, just as σ corresponds to the gradient of the distribution over s. Consequently, we cannot have N_λ specified sharply unless the gradient of the distribution over the phase ϕ_λ is singular. Here we see the incompatibility of the discrete with the continuous presented in its most striking form. The number of particles has to be counted. It is not a continuously variable number. Yet the analysis on which modern physical theory depends appears to represent this variable in a continuum. But this appearance is misleading, for the physical circumstances under which the number could be determined are incompatible with defining the corresponding phase. Thus sharp distributions over the integral system of possibilities for N_λ are possible, indeed the analysis ensures that only such distributions occur as the result of the mathematical operations that are said to represent the measurement of N_λ.

By means of the mathematical method, we seem to have assimilated the discrete operations of counting with the operations of physical measurement that lead us to continuous representation of the magnitudes measured. To represent quantum phenomena we have brought together counting and continuous measurement, and combined them in a single system in which we appear to have mixed up the discrete and the continuous. The latter is the natural

* See Akhiezer and Berestetsky, *Quantum Electrodynamics*, Ch. III, §16.

mode for mathematical analysis. In it, counting must appear as a jump or discontinuous transition.

Let us retrace our steps and consider an example of how we might interpolate distributions in the continuum of s, given the distribution over the lattice of integral values of s/a, represented by the vector $\sum_n a_n \delta(s-na)$ in function space. The probability that s takes the value na is $w_n - |a_n|^2$ provided $\sum_n |a_n|^2 - 1$. Let us displace the lattice in steps a/N on s, and in the displacement let the phase of the vector be changed by $2\pi/N$. Over the N lattices we have the distributions

$$a_n/N\, e^{i2\pi r/N}\, \delta(s-(n+r/N)a) \quad (r = 0, 1, 2, ..., (N-1)).$$

Imagine these superposed in function space; the resulting distribution is

$$\sum_n \sum_{r=0}^{N-1} \frac{a_n}{N} e^{i2\pi r/N} \delta\left\{ s - \left(n + \frac{r}{N}\right) a \right\}$$

$$= \sum_n \frac{a_n}{N} \sum_{r=0}^{N-1} e^{ir(2\pi-\sigma a)/N} \delta(s - na)$$

$$= \sum_{r=0}^{N-1} \frac{1}{N} e^{ir(2\pi-\sigma a)/N} \sum_n a_n \delta(s - na).$$

The first sum has a value significantly different from zero when N is large, only if $\sigma a \to 2\pi$. This combined distribution is therefore associated in the limit $N \to \infty$ with the singular distribution in σ-space, $\delta(\sigma - 2\pi/a)$. The corresponding continuous transform in s-space is of course $e^{i2\pi s/a}$. It is the eigenfunction corresponding to the singular value $2\pi/a$ of σ. The effect of superposing the distributions as we have done is to remove all trace of the original distribution over the lattice, except that $\sum_n a_n$ determines the amplitude of the new distribution; the amplitude of the latter does not depend on s at all. However, the phase does, and the periodic distribution exhibits the lattices from which we set out; they are selected by those values of s that correspond to the same phase (mod 2π). The phase increases progressively with s because we imposed the gradient of phase in displacing the lattice. If the phase gradient were imposed in the opposite sense the singular distribution in σ-space would be $\delta(\sigma + 2\pi/a)$, and $e^{-i2\pi s/a}$ in s-space. The lattice interval (a) in each case is reciprocally related to the singular value $(2\pi/a)$ of the transform variable.

In this continuous representation it is the phase that is associated with atomicity. A sharp value of phase gradient (σ) determines a lattice spacing (a) on s, each lattice being associated with a single value of the phase $(\mod 2\pi)$, and displaced lattices corresponding to the intermediate points of the continuum, being displaced in phase like the wave function for a harmonic wave of wavelength a. Conversely, a lattice spacing on σ-space and constant phase gradient for the distribution in continuous σ-space are determined by a sharp value of s.

Classically we expect to be able to specify all the relevant variables we introduce in our representations as sharply as we please, but it has long been recognised that the frequency of a train of waves takes a sharp value only for the ideal infinite train of simple harmonic waves with constant amplitude. The frequency and the time are reciprocally related as the variables σ and s considered above.

A possibility for existence that is imagined to persist or endure in time must be represented by a symbol that does not change with time, no matter how fine the representation. If with this possibility we attempt to associate the definite energy E_0, imagining that we may regard this association as effective with unlimited refinement in specifying the time, we are assuming that the mechanical action corresponding to E_0, viz. $E_0 t$, is continuously variable instead of restricted to integral multiples of h. Let us therefore mark off a lattice of instants $t = nh/E_0$, so that the action interval corresponding to the step in the lattice is h. If we attempted to represent time as counted in discrete intervals in this way, using a different lattice to deal with each particular instance, we should clearly be involved not merely with an awkward system of representation but with an incorrect one. For in giving the exact value E_0 to E we have selected a singular distribution in E-space. Since t is the variable conjugate to $E/\hbar$, the corresponding distribution on

t is $e^{i2\pi E_0 t/h}$, and the lattice of t-values that we were ready to introduce is exhibited only by the phase of this distribution which does
not correspond to timing at all. The persistence we imagine on
classical grounds without significant physical interaction is represented by the singular value of the energy only in a formal way.
We cannot establish the existence whose continuance is presumed
without changing the energy, that is, without changing the distribution by physical means so that it becomes one significant for
timing. Thus persistence is a quite inappropriate conception
applied to atomic existence.

We have spoken of a lattice as presenting a discrete system of
possibilities in the corresponding continuum. The points of the
continuum may be regarded as generated by continuous translation
of the lattice. One is therefore led to ask in what other respects, by
thinking of equivalent sets of elements in the continuum, we may
imagine limiting the possibilities of the continuous spatial relations
exhibited by different coordinate systems. Consider an axis
through a fixed point of physical space, and let there be a set of
n half-planes through the axis, dividing into equal sectors a sphere
centred at the point. Rotation of the figure through the angle
$2\pi/n$ about the axis turns the system of planes into coincidence
with itself. Here, rotation about the common axis serves with the
system of planes to generate equivalent systems, just as translation
serves to generate equivalent lattices. Whereas in the continuum
the longitude measured about the axis is continuous, in the system
analogous to the lattice there are only n distinct possibilities.

Instead of basing ourselves on axial symmetry to generate a
discrete system of possible longitudes, we may consider rotations
constituting a finite group of rotations about a point (the centre of
the sphere). A pair of integers, usually indicated by l and m, is
necessary to name the individual operations of the group.

Imagine that on the surface of a sphere of unit radius we know
the complex numbers, each of which is the value of a function at
one point of the sphere. As a physical example consider the value
of the electric force at a particular instant due to a small radio
antenna placed at the centre of the sphere. We choose the unit of

length (radius of the sphere) very large compared with the physical size of the antenna. The electric force, varying harmonically with the time, is represented by a complex number, following the practice of engineers who use the exponential function of imaginary argument in place of the trigonometric functions. The distribution of the electric force over the sphere is connected with the geometrical form of the antenna and the distribution of currents in it.

By means of longitude (ϕ) and co-latitude (θ), we may refer positions on the sphere to a polar axis and a half-plane it bounds, and we may regard the components of electric force distributed on the sphere as functions of θ and ϕ. With the aid of electromagnetic theory, however, we may dispense with considering separately the specifications of the individual electric force components, and treat only the distribution of two scalar functions of θ and ϕ. We shall assume that the antenna is so designed that only one of these functions need be considered to represent the field. When the antenna is turned round with respect to the polar axis and the meridian plane, the mathematical representation of the distribution of the scalar function $f(\theta, \phi)$ is altered by the rotation, but the physical distribution is unaltered. Let x, y, z be the Cartesian coordinates of the points of the sphere, so that

$$x = \sin\theta\cos\phi, \quad y = \sin\theta\sin\phi, \quad z = \cos\theta,$$

the z-axis coinciding with the polar axis. If $f(\theta, \phi)$ is equal to a homogeneous polynomial expression of degree l in x, y, and z, it may be expressed as a linear combination of the surface harmonics $Y_l^{(m)}$ of degree l, with $m = -l, -(l-1), ..., -1, 0, 1, ..., (l-1), l$. These are functions of θ and ϕ, which are unitary orthogonal and normalised over the sphere, i.e.

$$\int \overline{Y_l^{(m)}} \, Y_{l'}^{(m')} \sin\theta \, d\theta \, d\phi = \begin{cases} 0 & \text{if } m \neq m', \, l \neq l', \\ 1 & \text{if } m = m', \, l = l' \end{cases}.$$

We may therefore imagine a function space of $(2l+1)$-dimensions in which f is represented by a vector, or by the corresponding ray if f is normalised so that $\int |f|^2 \sin\theta \, d\theta \, d\phi = 1$. The components of the vector f, when this linear expansion in terms of harmonics of degree l is possible, are the coefficients attached to the corresponding basis vectors $Y_l^{(m)}$ for the generalised Fourier expansion. That is, we treat the basis vectors as a set of distinct logical possibilities. If $f(\theta, \phi)$ has the form that permits its representation in the space we have specified, it may represent the statistical combination of the set of $(2l+1)$ possibilities.

Being able to represent the radiation pattern of the antenna in the foregoing way depends on the structure of the antenna and the phases of the electric currents in its parts. Given that the pattern allows the representation of degree l, we require $(2l+1)$ complex numbers to specify it. The pattern is represented on a subspace of the entire function space which is required to represent a distribution on the surface of the sphere in general. When the antenna is rotated, its pattern is still represented by a ray in the $(2l+1)$-dimensional subspace spanned by the basis vectors of degree l.

The unitary transformation in function space corresponding to physical rotation takes each of the subspaces into itself, for the property that determines the degree of the polynomials in x, y, z required to represent the pattern is invariant with respect to rotation. Given a particular pattern of degree l we may, by continuous rotation with respect to a fixed reference system for θ and ϕ, pass through every possible pattern of degree l.

Thus, with respect to rotation, the connexion between the continuous and the discrete systems of possibility has a different structure from that between the continuum and the lattice established by continuous translation. The $(2l+1)$-dimensional subspace spanned by the surface harmonics of degree l resembles the subspace of periodic functions of a single variable, that can be represented by a finite Fourier sum with $(2l+1)$ complex coefficients.

Passing from the function space for the continuum to the discrete representations is the appropriate way to exhibit the mathematical structure of the system. However, by trying to proceed in the opposite direction we show ourselves how our simple ideas of atomicity and the discontinuous fall far short of enabling us to establish the connexion between the continuous and the discrete without the aid of the mathematical inventions. Nevertheless, in the process we see what these inventions are doing for us. For example, we have considered the unitary representation of a distribution on the surface of a sphere by means of surface harmonic basis vectors, and the unitary representation of a distribution in space by means of the Fourier components with respect to harmonic basis vectors. The former makes use of rotation about the centre of the sphere as the means of generating the continuum of possibilities, the latter employs translation. The family resemblance between these illustrates that while continuous transformations in ordinary space generate the continuum of possibilities from the discrete system, the representation in function space retains the structure of the discrete system in the appropriate transformations. Mathematically these considerations are approached through the theory of groups and their representations.*

The continuum presents us with continuous possibility. Atomic phenomena exhibit to us variables which take discrete values.† We have discussed as illustrative of this aspect of atomicity the relation of the lattice to the continuum. A distribution over the lattice is represented by a vector in a subspace of the space of functions of s. On Fourier transformation the corresponding distribution on σ is a periodic function with period $2\pi/a$. The discrete series on s is converted into a periodic distribution on σ, and the step interval on s determines the period on σ. Thus it appears that atomicity is associated with the continuous representation through the distributions of the phases of the complex numbers that

---

* G. Ya. Lyubarskii, *The Application of Group Theory in Physics*, translated from the Russian by S. Dedijer (London: Pergamon Press, 1960).

† We have spoken of discrete systems of possible values for certain physical variables as if the spectrum of possibilities were necessarily only discrete, whereas it is well known that the energy of an electron in the electric field of a proton has a discrete set of negative values corresponding to the binding of the electron in a hydrogen atom, and a continuous spectrum of positive values for the electron free to leave the vicinity of the nucleus. The continuous portion of this spectrum is associated with the unlimited space accessible to the electron. If we confined it to a box, we should have to replace the continuous spectrum with a discrete set of possible values which become more closely spaced the larger the box.

symbolise the logical possibilities of function space. Consider the simple case of the function $e^{i\sigma s}$. In classical terms, the product of the variables denotes an element of mechanical action divided by $\hbar$. For instance, if s denotes the classical space coordinate x, $\hbar\sigma$ denotes the corresponding classical momentum coordinate p_x. Whenever the product changes by 2π the change in action is h, and the function returns to its original value. Essentially this connexion in thought is introduced when we speak of a distribution as a wave. To say that the waves *must* be introduced on an equal footing with the particles to represent microphysical phenomena is a way of presenting some aspects of the essential logical structure that belongs mathematically to connecting the discrete with the continuous.

The foundations of our thinking about these matters are found in the mathematical generalisations that evolved from Fourier's series representation of a function of a real variable, in the concept of Hilbert space and generalisations of it* and in the analysis associated with it. These provide the apparatus for the representations used in quantum mechanics. Our purpose here is to understand what the successful formalism based on these mathematical inventions implies concerning the elementary assumptions from which we proceed in physical representation.

Our discussion is not confined to mathematical ideas since it involves also physical concepts. Nevertheless, the activity in which we are engaged resembles that which preceded the explanations of the differential calculus in terms of limiting processes and limits. Anyone accustomed to the algorithm through practice in use does not refer to these explanations. Indeed the calculus was used effectively in certain contexts before the logical exposition that we regard as satisfactory today had evolved. The mathematical inventions having been made in notation and the rules for operating with the signs, a learner could acquire skill in using it, and this experience served to determine his ways of speaking in using it. It

---

* P. A. M. Dirac, *The Principles of Quantum Mechanics*, 4th ed., p. 40 (Oxford, 1958). F. Villars in *Theoretical Physics in the Twentieth Century*, pp. 98 *et seq.* (New York: Interscience Publishers Inc., 1960).

may be suggested that we can observe a similar phenomenon in respect to quantum mechanics. The formalism itself supports much of the thinking, but it does not even yet fully enlighten us as to the metaphysical implications of its methods in representing physical phenomena. We have to examine it with the aid of other intellectual tools to reveal how its method derived in one way could be exhibited in another, and thus show its essential relations to ideas of discontinuity, atomic connexion, and their representation in continua. Mathematically, by means of the functional calculus of the theory of distributions,* quantum mechanics restores the derivative that we seemed to be deprived of, because the classical limiting processes cannot be applied to calculate it.† In this restoration, however, we are really deprived of the common instantaneous specification of a function (distribution) and of its derivative with respect to the time, for example, and the common punctual specification of a function and its derivative with respect to space.

It is well known, and we note here, that the distribution $ikf(k)$, where k is the variable conjugate to x and $f(k)$ is the Fourier transform of $\phi(x)$, transforms into the distribution $(d/dx)\,\phi(x)$ in x-space. We may therefore think of k as an operator on a testing function* in x-space. In that same space the operator x converts $\phi(x)$ into the simple product $x\phi(x)$. The operators x and ik, when combined, do not commute as they would if the signs were used as algebraic symbols for ordinary numbers.‡ The sense in which we regard x

---

* L. Schwartz, *Theorie des Distributions* (Paris: Hermann et Cie, 1951).

† Let $\phi(x)$ be the distribution whose derivative we wish to find. If

$$f(k) = \frac{\mathrm{I}}{2\pi}\int_{-\infty}^{\infty} \phi(x)\, e^{-ikx} dx,$$

then the derivative $\phi'(x)$ is given by

$$\phi'(x) = \int_{-\infty}^{\infty} ik\, f(k)\, e^{ikx} dk.$$

Both of these integrals can be computed even if $\phi(x)$ is discontinuous, provided that their convergence is provided for.

‡ For

$$(ik\, x - x\, ik)\,\phi(x) = \frac{d}{dx}\{x\phi(x)\} - x\frac{d}{dx}\,\phi(x) = \phi(x).$$

Thus $kx - xk = \mathrm{I}/i$.

and ik as operators is that they transform linearly the unitary vectors of Hilbert space, and can therefore be represented as matrices. Each element of such a matrix is labelled to correspond with the pair of basis vectors that define respectively the component of the operand vector and the component of the transformed vector connected by the element in the linear transformation.

It is surely remarkable that each pair of non-commuting operators consists first of the variable each of whose possible values names a basis vector in function space, and secondly of the operator which transforms an arbitrary vector in function space into its derivative with respect to the first variable in the sense of the theory of distributions. It is likewise noteworthy that the mathematical conceptions that free us from the restrictions implicit in the domain of continuous functions originated in the attempts to treat the discontinuities in terms of which one naturally represents switching operations in electric circuits.

In refining classical representation we encounter discontinuity and so are in mathematical difficulty with the limits on which we depend in applying the differential calculus. This formal difficulty is removed through the theory of distributions. What interest the physicist, however, are the ways in which atomicity limits the application of classical representation, and involves us with the new form of representation. For example, whereas classically we do not regard giving the physical variable x the particular value x_1 as the singular distribution $\delta(x - x_1)$ over the infinite continuum of x, we must so regard it in order to deal with the distribution over k-space derived by Fourier transformation. In this way we are compelled to regard the continuum in a different mathematical light.

It is shown in works on quantum mechanics that a statistical distribution according to the normal law in k-space with mean value k_0 and variance $(\Delta k)^2$ is transformed into the normal distribution in x-space with mean value determined by the gradient of phase in k-space and with variance $(\Delta x)^2 = (\Delta k)^{-2}$. We are thus led to interpret the commutation rule for the operators ik and x as the uncertainty principle $\Delta k \Delta x \geqslant \frac{1}{2}$, which couples minimum uncertainties Δk and Δx in the specification of k and x, respectively. The corresponding relation between frequency band width and precision in timing by means of an electrical signal is well known to engineers; it was forced on their attention by the filtering action of the electrical systems through which timing signals are transmitted.

While the uncertainty principle of quantum mechanics has usually been presented as characteristic of the limited precision allowed in specifying simultaneously the momentum and position of a moving particle, it is evident that the essential logical point of the principle inheres in the functional connexion between the distribution in the space of one variable and that of the other through Fourier transformation. This connexion is a mathematical one, and the analytical methods for treating it contribute in a striking way to representing on a continuum the superposition of discrete independent logical possibilities.

The uncertainty principle was originally explained in optical terms. The momentum of the particle being proportional to the wave number of the 'ψ-waves' cannot be specified precisely without compromising, due to diffraction, the precision with which the particle can be localised. Further, a physical explanation by means of a 'γ-ray microscope' was offered to explain how, in attempting to determine more precisely the position of an electron, for example, one necessarily communicates momentum to the electron by means of the radiation used to locate it. These explanations have served in the teaching of physics for a generation. But they do not resolve for us satisfactorily why the continuous representation of the 'motion of atoms' should be limited by this rule, except in one respect: the principle finds a role for Planck's constant in it. The motion of atoms is not motion in the classical sense, because of the existence of atoms of mechanical action.

If we wish to have continuous connexion in our representation of the 'motion' of the elementary particle it cannot be continuous connexion in the space in which we observe the events imagined as connected by motion. As we have remarked in two previous chapters, the essence of atomic connexion between events is that the connexion must be outside the space of the events. Thus the kind of connexion to which we had become accustomed in classical physics is ruled out, but we may establish systems of continuous connexion in function space. Here the continuity need not apply to the systems of independent logical possibilities, only to distributions over them.

The atom seems to put us in a situation with respect to representation not unlike that presented by the computing machine. Not only does it hold a finite number of significant digits, it allows us only finite representations with discrete systems of possibilities. For example, in computing, a function of a real variable is represented by a finite set of orthogonal polynomials linearly combined. Each polynomial suitably normalised corresponds to an independent logical possibility. The function can be regarded as a statistical combination of these possibilities. A corresponding limitation applies to physical measurement; it is imposed by the particular arrangement and procedure to make the measurement. We regard these limitations as practical only, not as intrinsic. By treating the finite representation of a variable number as an approximate one, we transcend these limitations to connect the finite representation with the continuum. Although our procedure is governed by certain general principles of physical measurement or numerical analysis, we have to work out in the particular case how they are to be applied. Accordingly, there is a substantial measure of unformalised activity in relating the numbers yielded by computing, or by physical measurement, to the continua of mathematical analysis.

We have to allow for the accretion of error in computing by 'rounding-off' or 'forcing' errors; and in the physical laboratory allow for physical effects that would affect the measurement because the causes of them are not under our control. But we are in no essential philosophical difficulty in understanding what we are doing in these processes. We are ready to accept the calculus of finite differences as a substitute for the differential calculus, and the finite sum in place of the integral. The number of significant digits carried by the machine is sufficiently great that the difference between the sum and the integral is reasonably small. The systems of discrete possibilities appropriate to atomic phenomena are much more limited than those of a digital computer, and this destroys the easy setting up of the correspondence between the discrete and the continuous that Bohr's correspondence principle exploited in the early days of quantum theory. But this is not the only respect in which atomicity restricts us. It deprives us of the kind of continuous connexions that characterise motion in classical mechanics and that played such a great part in the development of astronomy, and that we rely on, for instance, in using timing shutters to select particles with a particular speed in a beam of radiation.

In order to appreciate the effect of lattice representation, consider the display of spatial possibilities, first, using a rectangular grid and, secondly, using a grid based on polar coordinates. Unless we embed each of these in continuous space, connect the two by continuous transformations and use a rule of interpretation to introduce the limited resolving power of the grids when applying the continuous representation, we are faced with a very complicated problem in connecting the representations on the two 'lattices'. We should, indeed, have to rely on substantially the methods of quantum mechanics to handle the matter effectively, and, of course, that would really introduce the continuum again, but in a different way.

We rely formally on continuous connexion because we regard time as continuous, and, indeed, on account of the principle of relativity, are thereby required to regard space-time as continuous. The physical connexions between events in atomic phenomena exhibit the discontinuity and unity characteristic of atomic connexions, like moves on a chess-board. Quantum mechanics achieves continuous representations of systems of these atomic connexions, and this continuity is what serves us in relating atomic connexions to the continuous space-time of our classical intuition. If we wish to regard the representation of nature by quantum mechanics as something more than a formal artifice, we cannot escape altering our metaphysical approach to physical representation. We require non-classical concepts of atomic existence and of physical space-time, and a reasonable explanation of the way in which discontinuity and the other aspects of atomic phenomena are to be managed by continuous representation.

We depend essentially on mathematical inventions that have evolved contemporaneously with the physical theories which use them. Formal elaboration of the mathematical structure from the set of axioms congenial to the analyst does not serve to explain how the invention is relevant to the physical application, until theory presents forms that are well known to the experimenter. Whereas the relevance of the axiomatic structure should itself be explained. This is found especially in Dirac's *Quantum Mechanics*; but of course his interest was to set up the formal structure in a secure way. The discursive metaphysical comment in this book raises on occasion philosophical questions about physics, but since they are really irrelevant to the formal exposition itself, they are disposed of summarily. It is noteworthy that the concept of atomicity is not treated in these metaphysical preliminaries.*

The introduction of physical variables in conjugate pairs, one of which is represented as the derivative operator corresponding to the other, made it possible to use the forms of Hamiltonian mechanics in defining the dynamical systems which first served as atomic models. It did not take long, however, for the possibilities of the new method to be exploited in creating its own non-classical models. Since a physical variable is represented by a Hermitian operator on the function space, it was quite natural to consider as physical variables the operators by which the discrete indices distinguishing the components of a spinor, vector or tensor system can be changed. For example, if the wave involves a set of wave functions that are the components of a vector, an operator that interchanges components or forms linear combinations of them can serve to represent the effect of physical interaction by which, for instance, the polarisation of the wave is changed. An advantage of treating the transformations in this way is that thereby matrix

---

* The reader who appreciates concrete aids to thinking may find helpful the book by U. Fano L. Fano, *Basic Physics of Atoms and Molecules* (New York: John Wiley and Sons, Inc., 1959).

algebra may be employed to present the forms of physical interaction by simple symbols in a concise way. The choice of these
operators is governed by mathematical consideration of the relation
of the operators to transformations in space and time and in the
spaces of other physical variables. In this way the wave functions
for the electron were introduced by Pauli and Dirac, and many
of the theoretical inventions made in the theory of elementary
particles.

Our discussion in this chapter has attempted to exhibit the place
of the discrete and discontinuous and the place of the continuous
in our representations. Complex numbers are the vehicle of
atomicity because they can represent equivalent lattices on the
continuum. If we regard a distribution of a complex variable on
space and time through the physical model of a wave, we destroy
certain important logical elements of the proper representation,
namely those that we associate with the appearance of particles.
Both waves and particles belong essentially to classical physics.
The motion of a particle does not represent an atomic connexion
except as a discontinuous transfer of particle properties. The continuous motion of a particle does not represent this transfer at all
except in the classical context of a crude description. The wave, on
the other hand, represents the structure of a system of possible
atomic connexions between events in the wave field. The interference of waves superposes alternative logical possibilities represented by distributions of amplitude and phase, to produce a new
resultant distribution of amplitude and phase. From this we may
compute the distribution in space of the probability of individual
atomic existences in each of which there is established the connexion to a point of space where a physical interaction 'due to a
particle' takes place; or we may compute how the waves will be
diffracted by a physical system introduced for this purpose.

In our discussion we have intentionally avoided the mathematical details necessary for formal exposition that are found in the
standard works on quantum mechanics. Our purpose has been to
draw out the basic metaphysical reflexion that the mathematical
formality provides an apparatus for managing atomicity in the

context of continuity. To this end complex variables are the mechanism for counting periods that correspond to atomic steps. This is clearly enough exposed in the Argand diagram for a single complex variable. It is when we come to consider the generalisation of this in function space through unitary transformations that we lose track of what the algorithm does for us in relation to atomicity, because our attention is concentrated on mathematics. It is not enough to cling to the remarkable and powerful mathematical methods for intellectual support, it is necessary to see what they do for us in representation and to find a significant interpretation of atomicity by studying them. For the success of the methods testifies to their appropriateness for representing quantum phenomena. The mathematical logic has succeeded in embodying an effective concept of atomicity. To grasp it we must see why the continua are introduced at all. They are required on the one hand for analytical convenience, on the other hand they are essential for physical representation, because atoms have a statistical connexion with the world that we represent classically in continuous space and time.

CHAPTER VIII

INTERFERENCE AND ATOMIC CONNEXION

———————————⊃◆⊂———————————

The idea of an atom of any substance or physical quantity that can be measured requires us to replace the representation of continuous magnitude by a discrete set of possible values. It is the atomic step that takes us from one possible value to the next, like the single count in the serial operation of counting individuals. So an atom is not properly represented by a point: it is the connexion between two points and implies displacement in the corresponding space. If to each of the singular distributions denoted by these points there corresponds a stationary distribution in another space, the atom connects these stationary distributions. Thus we should say that the atom connects logical possibilities. Symbolically it may be regarded as transforming one into the other. If the distributions so connected are associated with separate events in space and time, the atom appears to move.

When we think of a distant source of atomic particles and form a beam by means of slits, we are concerned with a variable connexion, namely between the source and the field of events to which it is atomically connected. The variable connexion is one in the family of connexions we have in mind. For simplicity, we ignore the limited width of the beam of particles whenever we do not wish to represent the variations of intensity across the beam that we interpret as due to diffraction at the slits. We regard the intensity as uniform across the beam, so we represent the intensity as uniform across planes (yz) perpendicular to the direction (x) of the beam.

To associate the events with the distribution that we speak of as a wave, we survey the position of the source and the possible positions for the detector. This is done crudely compared with the refinement of wavelength measurements. Similarly, the timing arrangement is crude, so in effect we are dealing with the beam as if constituted of a group of waves covering a narrow range of frequency and wavelength.* That is, our representation by means of plane waves is an approximate one. We require the width of the beam to be many times the wavelength, and the distance to be many times the slit-width, and the latitude permitted in locating the source and the detector to be small compared with the slit-width and yet crude compared with the wavelength. We must adapt the experimental arrangement to these relative magnitudes in order that our simple representation by a plane wave shall be appropriate. This arrangement makes evident that the events are separated as required by the group velocity of the waves.

In this way we consider a simple family of atomic connexions. The individuals are uniformly distributed across the beam: they connect events that are separated in space and time in accordance with 'the velocity of the particles' which is equal to the group velocity of the waves because of the conditions under which we determine the displacement in time and space of the events connected by the atoms that constitute the beam. Let us now concentrate our attention on the phase distribution along x, assuming that we have indeed a singular value for the wave number, and thus overlooking the modulation corresponding to the group. The expression e^{iKx} represents not only the distribution of phase along x, it exhibits also a lattice of integral possible values of $S = Kx/2\pi$, via the structure of the transformations relating the discrete series to the continuum. So if K is equal to $p/\hbar$, where p is the momentum of the particles, the wave, in effect, implies the atomicity of action.

Let us now enlarge our representation to a two-dimensional spatial one. The plane wave $e^{iK_1x+iK_2y}$ defines lattices on the xy-plane with lattice intervals $2\pi/K_1$ and $2\pi/K_2$ parallel to x and y, respectively. Let $2\pi S = K_1x + K_2y$. Each lattice on xy corresponds to a lattice of integral possible values of S, namely, the integers n which are the sums of the integers specifying lattice points on xy. Lattices displaced parallel to each other on the xy-plane correspond to displaced lattices on S which are equivalent

* Contrast this with ignoring diffraction by the slit.

133

with respect to the discrete series of values n. K_1 and K_2 are the components of the wave vector K which is normal to the wave front in xy space. On a particular wave front, S takes a fixed value. Consider a transversal $x = $ constant making the angle θ with the wave fronts: the distance measured along this line is represented by y, and the wave on the transversal by $e^{iKy\sin\theta}$. On this distribution we superpose that corresponding to another wave of equal amplitude, but with its wave fronts inclined in the opposite sense to the first at the angle θ to the transversal. So the resulting distribution on y is

$$e^{iKy\sin\theta} + e^{-iKy\sin\theta} = 2\cos(Ky\sin\theta).$$

We produce this superposition by first splitting the beam symmetrically and examine the intensity distribution at a great distance, so that we have substantially two plane waves superposed at different inclinations to the plane across which we move the detector. The intensity distribution in the interference pattern is $4\cos^2(Ky\sin\theta)$. The wavelength for the interference fringe is $2\pi/2K\sin\theta = \lambda_f$. The change in phase corresponding to this displacement is $\pm\pi$ and takes place discontinuously at the node in the intensity distribution. The difference in phase between the two plane waves changes by 2π across one fringe, since for one wave the phase increases, while in the other it decreases by π.

Since $\hbar 2K\sin\theta$ is the range (Δp_y) of p_y, the transverse momentum corresponding to the two plane waves, we have shown that $\Delta p_y \lambda_f = h$. Accordingly, one fringe represents the limiting refinement in specifying y, so long as we interpret the wave vectors in the split wave system as limiting the momentum of the particles. Nevertheless, we can determine the distribution through the fringe, and can receive particles in a very much smaller range of y than λ_f. So we must accept that the amplitude of the resultant of the superposed waves is modulated in the y-direction, and that the response of the detector (such as a photographic plate, or a counter traversed across the fringes) is determined at a particular place by the local amplitude at the individual grain, or at the entrance to the counter.

The superposed wave system with the distribution

$$e^{iK_1 x}\cos(Ky\sin\theta)$$

must be resolved into its components with respect to the system of singular possibilities in space, $\delta(x-x_1)\,\delta(y-y_1)$, in order to compute the probability that the corresponding position will be occupied. The probability density at y_1 is proportional to

$$\cos^2(Ky_1\sin\theta);$$

it is the same as the intensity in the interference pattern.

But this way of looking at the matter is abstract and formalised. There is no effect detectable at $y = y_1$, unless there is something in space to register the effect that we speak of as the detection of a particle. To represent this process we analyse the plane* wave $e^{iK_1 x}$ into a system of ingoing and outgoing spherical waves centred on a suitably chosen point of the small piece of matter where the detection can take place. The possibility of the physical effect that constitutes the basis of detection is represented by transforming the spherical wave system, so that the outgoing waves no longer match the ingoing waves in amplitude and phase. So the presence of the detecting system is represented by superposing on the incident wave system, the waves denoting the possibility that, the detector having undergone the atomic transition that constitutes detection, a particle has been removed from the beam. Using the appropriate physical model we can compute the probability of the particular atomic transition by the methods of quantum mechanics. If we do not wish to particularise the detection process in detail, we may associate with the detector an effective cross-section (σ, an area) which is numerically equal to the probability of the transition to be detected occurring when the intensity of the beam is one particle per unit area.

If the detecting elements are sufficiently numerous across the beam, and uniformly spaced on the average, the distribution of intensity of the radiation, as measured by the distribution of

---

* We may replace the modulated wave by an unmodulated one if λ_f is very much larger than the range of interaction between the atomic system of the detector and the particle to be detected.

detection events, will approximate that yielded by the simple formula $N\sigma A^2$, where A is the amplitude of the incident wave, and N is the number of detectors per unit area.

When we speak about the matter in this way, we have essentially changed our attitude from that we adopted when we wished to say that the particle arrived and by its presence caused the physical effect by which it was detected. Nevertheless, within the limits permitted in classical descriptions, this way corresponds statistically with physical thinking: the particle is unlikely to produce physical effects where it has little chance of reaching, because the amplitude of the wave is very small there.

In the imagined observation of interference we connect the beam of particles with the distribution of individual counts along y. The events counted, we may remind ourselves, are crudely located compared with the wavelength of the radiation, because we are dealing with actual things. Yet in the mathematical representation, we use the ideal location at a point. Clearly that localisation represented by the singular $\delta(x - x_1)\,\delta(y - y_1)$ is a mathematical fiction, because any system we use to detect a single particle has some spatial extension. So long as this is small compared with the wavelength, the δ-distribution serves, and it has the great convenience that we do not need to specify the particular physical object that serves to show that it has 'interacted with a particle'.

We have remarked that the single fringe corresponds to the atom of action. So far as the distribution over y is concerned, the fringe represents a single spatial possibility for the particle in the absence of matter that can diffract the waves, or that forms with the radiation a system that undergoes atomic transitions the result of which can be observed or is recorded. That we are able to discriminate the intensity distribution in detail over the fringes, surely shows that the distribution of intensity reflects how the interference was produced. The structure of the interference pattern tells us about the system used to produce it. Given harmonic waves, the wavelength λ determines the scale of the pattern for a given interference arrangement.

The distribution we observe depends on the geometrical disposition of the arrangement to split the beam. We compute the intensity distribution along y by the wave calculation, but in terms of the momentum and direction of motion of the particles, this calculation is equivalent to superposing two distributions with well-defined momentum in slightly different directions. This way of putting the matter treats the alternatives presented by the two slits of the interferometer as distinct, as indeed they are, but it requires us to superpose the distributions arising from them to find the distribution that represents the system of possible atomic existences. It is a distribution both in amplitude and in phase. The former is exhibited by counting many particles as the pattern is traversed by the counter; but this measurement does not reveal the phase distribution. There is a gradient of phase in the x-direction: it corresponds to displacement parallel to x of the distribution across planes at right angles to x. This progressive change of phase represents the system of atomic existences as transferring momentum and therefore resembling a shower of particles. Across the interference pattern, however, we find standing waves. In them the phase distribution is discontinuous, and if we allowed the radiation to pass through a hole in the receiving screen, exactly placed to allow two adjacent fringes to pass through the screen, we should split the beam again on the far side of the screen.

The foregoing considerations may be extended to the superposition of a system of coherent plane waves, as from a grating, incident at different angles on the plane where detection takes place. We should find that since the fringe spacing is determined mainly by the difference between the extreme values of the components of the wave vectors parallel to the detector plane, that spacing exhibits the existence of the atom of action in substantially the same way as was shown for the simple system.

The actual intensity distribution in the fringe system of course depends on the whole wave system. The intensity distribution is in a certain sense, therefore, a picture of the source. It is a statistically derived picture computed by means of waves.

So long as our attention is directed to the atomic connexion between the events in which the effects occur that we associate with the departure and arrival of a particle, we avoid the difficulties in conceiving how the particle moves between them. For in dealing with interference and diffraction, we no longer regard the connexion as established by the rectilinear motion of the particle between the events. On the other hand, when we represent the interference phenomenon by means of waves, although our classical practice leads us to attempt it, we do not regard them as waves in a physical field. We use a different conception of the connexion between possible events. We are concerned with connecting distributions over systems of the logical possibilities, and we interpret and represent the latter in terms of the continuous spatial possibilities associated with physical measurement. We do this because of our predilection in favour of continuous change in physical space; but the atoms and discontinuity are there nevertheless. It is true that quantum mechanics returns to continuity in function space to represent the evolution of change in a physical system; but the continuous change does not take place in the space of the variables we measure and for which we may find only discrete values. The change is represented by a changing distribution over the discrete possibilities. Even when a variable is not restricted to discrete values, the change is still represented by means of a changing distribution over the whole space of the variable.

The effect of this method is that the connexions between real events are represented as belonging to systems of connexions. By means of these systems we can assign probabilities to the various observable alternatives. Thus the most important aspect of an atomic connexion is preserved. It has no internal structure. Nevertheless, the system to which we conceive it as belonging does have an internal structure. In describing this the formal waves are relevant. They represent how distributions may be transformed in space and time. The plane wave e^{iKx} transforms a uniform equiphase distribution across the yz-plane, for example, by displacing it parallel to x.

By concentrating attention on the independent logical possibilities—which may themselves be distributions, or waves—we separate the continuity associated with continuous representation from the discontinuity and discreteness associated with atomicity.

We should note, of course, that marking a particular logical possibility as occupied merely represents the logical context in which existence is possible, for physical existence stands outside this system; existence refers to the supports on which the formal system relies but which it does not represent. An electron imagined in an atom is associated with one of the atomic states, or, in a changing state of affairs, with a distribution over several states. During the atomic transition that results in the emission of a photon, it is the distribution over the states that changes, and as Heisenberg first understood, the electromagnetic properties of the atom determining its radiation field depend on both the initial and the final states. So long as we think of states as logical possibilities which may be occupied or not, and the superposition of states as a logical combination of possibilities in which the use of complex variables is understood as a necessary bridge between discrete and continuous representation, we are not perplexed by the question 'how can the electron be in two states at once?'. For this is resolved when the atomic transition has been accomplished.

The foregoing discussion serves to remind us that the connexion between the source of the radiation and the field of events in which detection may take place is by means of waves. They stand for possibilities in the sense of a distribution over function space. Superposed waves stand for the joint presentation of alternative possibilities; they are statistically independent only when they are mutually orthogonal in function space like the basis vectors of a complete system. If they are not statistically independent, interference is produced in the superposition. The existence of phase distributions is exhibited by superposing waves representing coherent logical alternatives, and measuring intensity distributions over an appropriate space. Nevertheless, the phase distribution is associated with our continuous representations because the connexion that we speak of classically as the motion of a particle between the individual events is in fact an atomic connexion. The waves do not constitute this connexion at all. They are essentially outside it for they represent distributions of amplitude and phase over the field of events.

In the preceding chapter we discussed how in representing a system of discrete possibilities by means of the continuum we resort to the domain of complex numbers. The particular system, that treats a position coordinate of a particle and the corresponding momentum coordinate divided by $\hbar$ as conjugate variables for Fourier transformation, exhibits Planck's constant as an atomic unit and thereby limits classical representation. This atomic unit determines only the scale of atomic phenomena. It does not touch the important logical aspects of our atomic concept in respect to motion, of atomic connexion, and of our need to look at the continuum through the apparatus of Hilbert space so as to make it possible to connect points of physical spaces by continuous connexions that do not lie in the physical space at all. The energy jump of an atom in emitting a photon is a discontinuous process: we cannot speak of the energy as changing in a continuous process represented in physical time.

We have to give up continuous connexions in the spaces we use to represent physical measurements in the classical sense. The variables we determine by physical measurements are not to be represented as singular possibilities in the continuum; they are distributions over the continuum. Accordingly they can be represented by superposing other distributions.

In interference we are concerned with a coherent splitting of a distribution of amplitude and phase in a part of physical space and the subsequent reunion of the transformed partial distributions. The distribution is represented by a vector in function space: it is decomposed into components in the splitting, and the transformed components are added when the reunion takes place. The fact that we have to deal with coherent decomposition and combination shows that we are operating merely in the field of possibility, and that we introduce no process that could bring one of the atomic existences to light. The partial distributions are split from each other in one region of the space of the physical variable and combined in another region after a transfer process determined by the dynamical law governing such transformations in the space of the variable in question. Thus in the simplest case, we take two subspaces of the function space (corresponding to the non-vanishing components of the vectors produced by the splitting), imagine vectors in them to be transformed, and consider the resultant arising from superposing the distributions over the derived subspaces. Interference takes place only when the subspaces in which the distributions differ significantly from zero overlap, and there both have non-vanishing components.

This way of looking at interference releases us from the intuitive responses we make in using classical representations of waves or motion of particles, because it takes us out of physical space as the space of the representation. For this reason we are better able to analyse what is going on in it.

The atomic particle must be kept out of the representation between the source and the detector for it connects them in one jump. Physical waves are also kept out of the picture unless we have a physical magnitude that could be measured as a classical variable and represented in the appropriate continuous space adjoined to space and time, like the electric force in an electromagnetic wave. For the representation is statistical in effect; it appears to get inside the atomic connexion but this appearance is due to our use of continua in representing possibility for the purpose of computing the statistics.

Coherence is relevant to the integrity of the atomic connexion. Any output from the statistical algorithm is achieved by eliminating the representation of phase; this elimination corresponds to the effect of the physical interactions by which the system could be observed between source and detector. These being atomic interactions, we have no means to attach a meaningful phase to the state vector in the transformation that represents the interaction. For phase distinguishes differences in the continuum. To measure it we depend on interference which in turn is exhibited only between coherent waves—in a well-defined relative phase.

The production of interference phenomena with ordinary light, when the individual atoms of the source spontaneously emit light incoherently with respect to each other, is accounted for in quantum representation by the waves that we associate with the individual photon.* So we are quite baffled in trying to understand classically how the statistics of photons in the interference pattern are to be explained, for we have no means of connecting the waves with the motion of the individual photon as processes in space and time. On the other hand, when we have a source consisting of coherent atomic radiators, the phase, being governed by stimulated emission as in a maser, is well defined no matter which particular atoms undergo electronic transitions to release energy.† Accordingly we have here an electromagnetic field of optical frequency like the fields of much lower frequency produced by a radio transmitter. Provided that the intensity of the field is sufficiently great that we may ignore the fluctuations inevitable with quantised energy, it may be treated by the classical methods applied to radio waves.

According to the quantum theory of boson fields, the uncertainty in the number of particles associated with a harmonic wave of definite frequency and wave vector is coupled with the uncertainty of the phase of the wave by the same commutation rule that connects wave number and the length of the train of harmonic waves. Precisely specified phase is incompatible with a meaningful specification of the number of particles with the corresponding energy and momentum in the field. As the number of particles is made larger and larger the phase can be better specified and, of course, the statistical fluctuations in the numbers of particles are correspondingly reduced relatively to the mean numbers. So the physical wave is a limiting conception that may be attached to classical modes of thinking when the number of one kind of particle in the field is large under conditions of coherent excitation.

Whatever the technical complexities associated with computing interference, due to polarisation, and phase shifts and the amplitude distributions that are known as form factors, one simple fact stands out. The interference pattern is a projection of the arrangement that produces it. As the means of studying a microphysical

* A. I. Akhiezer and V. B. Berestetsky, *Quantum Electrodynamics*, ch. 1, translated for the United States Atomic Energy Commission, Oak Ridge, Tenn. (original Moskva, 1953).

† A. L. Schawlow, 'Optical Masers', *Sci. Amer.* **204**, no. 6 (June 1961).

system we use a particular radiation that may be restricted as to wavelength and even polarised. It determines principally the scale of the pattern relative to the geometrical disposition of the apparatus we use, and it also determines certain aspects of the particular projection realised. Nevertheless, since the pattern is a distribution of individual atomic effects each of which bespeaks an atomic connexion by means of the radiation, the pattern is essentially a statistical projection subject to statistical irregularity. However, just as the iteration of a stochastic process reveals eventually a well-developed trend that images the probability distribution over the possibilities for one step, so as the number of counts is increased, the interference pattern tends to a smooth distribution that is connected statistically with the quantum mechanical model of the physical system that gives rise to the pattern with the radiation used.

By means of X-rays and the electron microscope we produce effective pictures of the atomic structure of matter, and by means of the high-energy machines we produce pictures of nuclear structure. We use quantum mechanics to apply these pictures, but nevertheless in principle they are as valid pictures of physical reality as those we make with the classical means. Since we make representations in space, we are able to present these pictures as if they were classical pictures. The fact that this is not how we use them in physics is irrelevant in the context of a descriptive account of what has been achieved by physical research when that account is presented to learners without the experience to understand better. We do well to remind ourselves that the pictures of reality presented by sight and on which we rely in everyday living are in fact produced by the same means. Photons are received by nerve endings on the retina and from this statistical input to the physiological system that has evolved to handle them in the context of life, we experience reality in the terms we have learned to interpret through the representations associated with ordinary language. The processes by which smoothing of statistical input is achieved cannot in the nature of things be like those that the mathematician invents using continuous representation. It seems much more reasonable to guess that the smoothing processes have more in common with what happens in computing when we force the representation of the result of an arithmetic operation to a finite number of digits. Thus it is by a process of ignoring details that significance emerges. Since the biologically significant is recognised in some biological function, the fact of statistical fluctuation is irrelevant with respect to the system of possible forms to which the particular biological function responds effectively. This resembles in a remarkable way the response of a physical atom to changes in its environment. It has only limited possibility; responses to changes that can be discriminated by other means show only in the statistical distribution of atomic transitions over the possibilities open to the atomic system.

Thus it appears that escaping from the metaphysical commitments of ordinary language, and the methods of continuous representation by which they involve us in irrelevant possibilities, is perhaps the outstanding intellectual problem of modern science and of understanding not only science, but how we experience life as a whole.

The continua of classical physics serve important and useful purposes, but their use depends physically on the statistical regularities appropriate to large numbers. So the metaphysical attitudes on which classical physics was grounded really exhibit the primitive state of scientific knowledge at the time modern physics began. Only when sensitive detectors of atomic events were invented did it become obvious that these events are not merely incidental to observing phenomena: the world is changed by the physical consequences of their happening. Atomic interaction in the world is the mechanism of all real change.

In an event implying detection of a single atomic particle, the physicist depends on some arrangement that amplifies the physical effect to be measured and registered by a multiplicative process. Either the primary detection itself releases a substantial energy quantum which is degraded into a large number of quanta each of much smaller energy and can therefore be readily detected, or the primary detection event triggers the release of energy in a system such as a Geiger counter. Whatever ingenious device is invented to provide reliable detection, its role in relation to the phenomenon being observed is to register individual events belonging to a class of events selected by the device. The physical mechanism of the device itself is not under study when it is used in this way.

In detection a place in space-time is associated with at least one transformation in another space. There are sometimes classical measuring processes associated with detection, as, for instance, deflexion of charged particles by electric and magnetic fields, or of atomic or nuclear magnets; and we may note also, timing devices and coincidence detectors, and selectors such as pulse height analysers. With all of these it is significant that classical physics is used to support the measurement, for classical ideas of existence are adequate. These remarks apply also to the use of wave concepts in measurements that depend on diffraction, change of polarisation and so on. Since statistically valid results are required, large numbers of particles are detected and counted. In this way and by operating far from the limits set by the uncertainty principle, the experimenter encounters no theoretical obstacle in the measurement.

The atomic event requires a discrete finite change in the detecting system. Passage to the classical idea of measurement is through an imagined continuous process, for example, a continuous electric current (statistical average of the rate of collecting electronic charges) or a continuous force (statistical average of the rate of momentum change composed of individual impulses). Classical measuring apparatus operates on such statistical averages, and the change that is so obvious when we deal with the counting process is overlooked. For example, in using a Faraday cage with an electrometer, the latter is reset at zero in a process that is not part of the physical phenomenon being studied, or the current goes to ground through a current meter in which case there is no cumulative effect due to the arrival of charged particles in the Faraday cage because of the large electrical capacity of the earth.

In a counter, resetting is accomplished by electronic means at high speed and the pulse of electric current that signals the atomic event triggers the counting mechanism. The latter consists of a circuit that cyclically passes on pulses at a much reduced rate to a register on which the count is displayed. In the counting process, the progressive accumulation of counts occurs in this register; all of the circuitry between it and the primary detector of atomic events is reset automatically so as to repeat in each part the appropriate signal to transmit counts to the register.

This physical machine when functioning properly operates without statistical vagaries that are physically significant, because the ratio of the signal voltages to the voltages due to noise is kept large. Unavoidable statistical fluctuations in the count may be observed, however, due to random physical causes affecting the primary detector. How this apparatus functions as a counter is explained classically. Nevertheless, the measurement achieved by means of it shows that it differs from the classical physical measurement in an important respect, namely: there is a physical difference in the measuring apparatus after the measurement whereas in a classical measurement of a continuous variable by a meter the physical situation is not changed by the measurement and the indicator returns to zero when the cause of its indication is removed. The counter therefore exhibits the basic difference in principle between the classical concept of measurement without change in the physical situation, and the concept that has been associated with the measurement of quantum processes according to which the physical situation does change as the result of measurement. It is quite obvious that the change we have been considering is not a microphysical one, but that fact reinforces criticism of the concept of measurement of a continuous variable that ignores the counting process required in practice—in the marking of measuring scales, for instance—to introduce numbers into measuring.

CHAPTER IX

THE NEW AETHER

The world of our everyday life is presented to us in forms we have learned to recognise and to name. We have inherited a concept of the Earth on which we live, of its motion round the Sun, and of vast empty space that separates the Solar System from the nearest star, and so on. There have come down to us numerous inventions, tools, instruments and techniques that aid our thinking and enable us to represent the motion of the heavenly bodies; such representations have great practical importance. Euclidean space and Newtonian time have played such a dominant role in the evolving adaptation of man to his physical environment that they have become integrated in the logic of ordinary language and our use of it. From our experience of the objects we handle, we learn primitive ideas of their properties and the effects of our acting on them. This process is accelerated through formal instruction in physics and we are led to think not merely of the machines we have made operating on our environment to change it, but of the processes which we imagine must be going on in the world beyond our control, but nevertheless affecting us. How we should represent these processes seemed at one time a matter for *a priori* judgment. The evolution of physics has shown us that new methods are developed as we gain fresh experience and establish new knowledge by experiments and by theorising about them.

In the nineteenth century the wave theory of light was first elaborated by means of the mechanical model that represented light as distortional waves in an elastic solid. In the same century this picture was superseded by the electromagnetic theory of light, and the elastic waves were replaced by waves of electric and magnetic force in 'empty space'. The elastic solid aether of Fresnel became the electromagnetic aether of Faraday and Maxwell.

In this representation the essential substance is the electromagnetic field which we imagine specified at each point of space and changing with time according to the differential equations of Maxwell's theory. We can measure the strength of the electric and magnetic fields in radio waves by the techniques of the electrical laboratory. In an electromagnetic field of optical frequency we infer the field strength from a measurement of the energy flux using Poynting's theorem. The recent invention of the optical maser, however, and the production of beams of coherent optical radiation, encourages the expectation of great changes in the experimental techniques to measure the field in electromagnetic optics. For when the field is sufficiently intense the statistical fluctuations characteristic of the presence of particles are feeble relative to the field magnitudes to be measured, viz. amplitude and phase. In principle, our classical picture of field vectors, specified precisely in space and time, and that can be measured like any other physical quantities, constitutes the electromagnetic aether.

But is the electromagnetic field a substance? It is true it is not matter in the naïve sense of ordinary language—although some men once wanted to regard the aether as a material substance of which the electric and magnetic fields represented its physical state. The field plays the role of substance, however, in that it attaches to space-time the spaces of possible values of the field vectors.* The empty space-time of our representation corresponds to unspecified field vectors, like a variable x in algebra that has not yet been given a value. Whereas a vacuum, that is, with nothing physical going on in it, is represented classically by zero field strength everywhere. It is important to distinguish between a representation with respect to which some function or other of space and time is left variable, unspecified, and one in which the function vanishes everywhere. These are logically two different representations.

The substantiality of the electromagnetic aether in the sense of Newtonian relativity was abolished by the Michelson–Morley experiment. Since the advent of the special theory of relativity

* See *On Understanding Physics*, ch. v.

and the formulation of physically compatible transformations of space-time coordinates and of electromagnetic field components, associating the field vectors with space-time has to be understood in a new way, even from the point of view of the classical continua. Formally this representation plays its part within the limits imposed by the laws of quantum theory.

The classical electromagnetic field can represent the storage and transmission of the energy and the momentum transferred by the field to objects on which electromagnetic radiation is incident. The field is connected to physical objects whose electrical properties couple them with the field—generators, magnets, radio transmitters and receivers, for example, and so on through the apparatus of electrical and electronic, including quantum electronic, technology. Through its intervention, the electrical interaction of such objects separated in space is supported locally, so that the conservation of energy and also of momentum operate consistently with the special principle of relativity. But it does not correctly represent interaction between these objects and atoms, or between atoms and the field. The classical field exhibits none of the particle properties associated with wave propagation that we have learned to treat by means of quantum mechanics in order to represent the quantisation of energy and momentum in a field. So in its turn, the electromagnetic field, conceived classically, has outlived its usefulness except for engineering. It is replaced today by quantum fields each governed by its own field equations. The representation is much more elaborate than that of the electromagnetic field. Whereas the latter is specified by means of a six-vector (or antisymmetrical tensor) or by means of the four-vector electromagnetic potential, the quantum fields are specified in turn by scalar, spinor, vector, tensor and other mathematical quantities, each of which has its appropriate laws of transformation compatible with the space-time transformations corresponding to translation, rotation and so on.

The mathematical theory establishes a well-knit net of connexions for representing the classical mechanical properties of the fields. Each conservation theorem, for example, momentum, energy, or angular momentum, is connected with the invariance of the field when subjected to the appropriate infinitesimal transformation or operation in space-time: displacement of the origin of the space coordinates, displacement of the origin of time, and rotation of Cartesian space axes, respectively. This connexion between conservation theorems and groups of transformations was brought to light by Emmy Noether over fifty years ago. Thus mathematical thinking has guided the choices of the forms used in theoretical physics in a fundamental way.

By means of the formalism of quantum theory, the Fourier analysis of each field is used to develop means for representing the field as an assembly of particles—photons, electrons, and so on— and this same method is applied to represent elastic waves in solids and liquids by means of phonons. The energy and momentum of these particles are related to the frequency and wave number of the corresponding waves by the well-known quantum relations. The interaction of one of the fields with a particular physical system is represented by reducing the number of particles in the appropriate momentum states in that field and imparting the corresponding energy and momentum to the system ('absorption of particles by the system'), or by increasing the number of particles in the field, corresponding to removing energy and momentum from the system ('emission of particles by the system'). The field theory determines the probability of these exchanges according to the form of the interaction between the field and the physical system considered. The latter must be localised, at least crudely, if the mechanical effects on it are to be observed in fact. But not only are the fields coupled to physical systems we can observe, they are coupled to each other. In these mutual interactions, particles are annihilated and others created. The mathematical apparatus for representing such changes is established by regarding the field symbols not as mere magnitudes attached to events in space-time, but as accounting operators on the 'books' for the entire system of interacting fields. A creation operator increases by unity the number of particles in the appropriate state of the field concerned and changes the momentum for the entire system. Other physical properties of the system such as electric charge, even Coulomb field,* must likewise be affected. An annihilation operator has the opposite effect.

In the interactions between fields, we consider imaginary values possible for the momenta of the particles created. These correspond to evanescent waves localised near their source. On account of the relativistic connexion between energy, momentum and mass (or, in wave terms, between frequency, wave number and cut-off frequency), the energy to be associated with the particle in a mode represented by an evanescent wave is less than the rest energy of the free particle. It is then called a virtual particle. It connects the imagined event in which it was created, with the imagined event in which it disappeared. Indeed, the connexions between events by means of particles of the interacting fields constitutes the imagined structure in the model. Nevertheless, just as the classical field is a symbol to represent the possibility of observable effects when the proper physical devices are placed in space to exhibit them, so also these connexions deal only with possibility. Because of the statistical implications of quantum mechanics, the symbol for the field is applied in a way quite different from the classical one, although it exhibits certain formal resemblances to it.

Since interaction between fields consists in changing the numbers of particles they represent, a non-vanishing field of one type can act on zero field of another type. A better way of expressing this thought is to say that the fields are coupled to each other and can excite each other—like the stimulation of different modes in an empty wave guide by means of an antenna inserted in the guide.

It is suggestive to compare the coupling of the fields in quantum field theory with the excitation of different characteristic modes of wave propagation in a guide. The antenna in the guide introduces a conducting surface on which electric currents may flow. These currents are sources of both propagated modes and evanescent modes with amplitudes that depend on the position and orientation of the antenna in the guide; currents are likewise induced by waves incident on the antenna. We may regard the current system as

* Cf. P. A. M. Dirac, *Canad. J. Physics*, **33**, 650 (1955).

determined nearly enough by the condition that the field is bounded by the surface of the antenna, which is therefore represented as a surface across which discontinuous changes in the electromagnetic field occur. It is this boundary that imposes the coupling of the modes.

When we think of the coupling of fields, we might be tempted to regard each field as a distribution of sources of the other. But this is not correct. It is only when one field undergoes a change—an atomic transition—that it induces changes in other fields. The field change consists of the emission and/or absorption of the corresponding field quanta or particles. It is therefore not at all unreasonable to think of the imagined events constituting the stochastic process of field interaction as standing for physical possibilities that might be realised. But they must not be treated as realised when we compute the statistics of the end results of interaction under conditions that do not permit their realisation. Such interpolation alters the physical situation by creating a different phenomenon.

To speak of the number of particles existing is not correct, except when it is understood that the number quoted is a statistical average computed according to the appropriate algorithm of field theory. It is therefore subject to fluctuations. Here we seem to stand on very strange ground because our ordinary conception of physical existence is associated with the persistence of the things we meet in our ordinary experience, unless they happen to be destroyed in a process that we can observe, such as fire or some other chemical transformation. The physical properties of ordinary matter can be changed by heat, and chemical substance can be altered, but the energy is not destroyed in these processes. Energy endures continuously. According to the modern theory of elementary particles, however, we resign from our classical conception of continuous motion and the localisation of energy is not sharp, nor can energy changes be timed instantaneously in the fields. Indeed, the word 'particle' in this context tends to mislead us.

In experiments to investigate the interactions between the various elementary particles, we direct on a piece of matter a beam of particles endowed with great kinetic energy by means of an accelerator, or liberated in a nuclear reaction initiated by accelerator bombardment. The products issuing from the piece of matter in the beam are studied sometimes by means of processes that are adequately represented by classical methods, because the fields of force acting on them are not given structure on too fine a spatio-temporal scale. Sometimes they are studied by means of processes that do introduce at some distance from the first target, secondary nuclear, or atomic, interactions which distinguish the products in respect to type of radiation, energy and polarisation, in accordance with quantum mechanical methods of representation. Crudely expressed, we may say that the word 'particle' finds its proper context in referring to the incoming bullets that import energy to initiate the physical effects to be studied and in referring to the outgoing bullets that can be counted. The particles are not merely atoms that transport quanta of energy and momentum, mass and electric charge in the appropriate field. They are in fact atomic connexions between events in space-time, which is represented as a continuum in the mathematical theory.

When we try, as we do classically, to imagine the particle as something travelling through space like a bullet, as if it transported substance by continuous motion, we are putting the substance in the wrong place. The particle is localised only where it is created, where it is annihilated, or where it interacts with a localised field of force, and in each of these instances the localisation is crude. It is not punctual as it would be in classical physics. The field theory serves to manage the necessary statistical distributions of the incoming and outgoing radiations and their connexion with our physical model.

We imagine particles ejected from the source and we use essentially the classical idea of their existence in the beam of radiation because the beam can be treated for many purposes as a classical agent on detectors moved into it. Likewise, we can project back to the target from the devices by which we detect the outgoing radiation. The incoming beam and the direction of the outgoing particles intersect at the target. If not, we must imagine that we have detected a secondary (or tertiary, etc.) radiation coming from some matter other than the target. Thus we have existences supported by the apparatus for producing the incident radiation and by our activities in relation to them, and we are likewise assured of existences counted in the detector of outgoing radiation. These assemblies of existences are connected statistically at the target. We imagine an incoming particle, say, a proton, interacting with a neutron, for example, in a nucleus of the target material. The grounds for thinking in this way are found in various experimental observations that have been made. For instance, the physicist may compare the observed results from a deuterium target with those from a target of ordinary hydrogen. The former contains neutrons, the latter does not. There is no end to the variety and ingenuity of the methods an experimenter can invent and is ready to interpret in terms of current theory.

The spatial range of the high-energy interaction of the incoming proton with a neutron in the target is effectively less than 10^{-12} cm. In the centre-of-mass system of coordinates for the interacting particles we are dealing with very short wavelengths. The nucleon fields are coupled to meson and hyperon fields. By means of field theory, one can compute nuclear cross-sections for each of the various processes in which elementary particles are produced, and one can imagine different stochastic series of connexions inside the small sphere within which the forces are effective. Some of these may contribute to the particular reaction studied; the corresponding outgoing fields are combined coherently to determine the angular distribution, intensity and polarisation of the outgoing radiation. The stochastic connexions introduce virtual particles. Are we to say that they exist? Or, are we to say that this is merely a picturesque way of describing the mathematical form of connexion (S-matrix) between the incoming beam and the outgoing radiation that is detected?

Clearly these are questions in answering which we should merely express how we are disposed to regard the physical interpretation of the mathematical theory. For the words 'exist' and 'existence' refer not to abstraction but to real life. Their application involves how we commit ourselves in action with respect to the objects in question. Virtual particles are certainly not to be regarded like the things that form part of our ordinary environment. But neither are electrons, radioactive nuclei, or mesons. The last-named are noteworthy because of their extremely brief existence. Yet physicists have no doubt of their existence relative to the apparatus and techniques by which they produce them, detect them, deflect them in their passage through magnetic fields, and scatter them, and so on. Thus the sense in which virtual mesons or other particles will be said to exist will be established by the development of experimental methods to investigate them with the support of the theory on which we have learned to rely.

We must resist the temptation to think of the difference between an idea and the reality it represents as an occult one. The difference lies in our being involved with our representation in different ways. Existence is not a property like hardness or colour or weight, although experience of any one of these may dispose us to assert an existence. Asserting an existence bears some resemblance to asserting that a calculation is correct. It is the signal that, at least for the present, checking the calculation is over. But it does not guarantee that subsequent experience in using the result of the calculation may not induce us to question its correctness.

These reflexions about the existence of physical processes imagined in applying quantum field theory serve to emphasise that this word refers to how we go on in using our representation. The fact that the future use is not yet here (cf. Augustine on 'time') seems to stand opposed to our metaphysical inclination to attribute some quality now to the object whose existence we assert. One is reminded of Hume's allusion to the greater vividness of the images of real things we can see and handle compared with things merely imagined. The impression of vividness is real enough in much of our experience, but is this the criterion? Clearly not. We judge this matter in the light of further experience. Rarely do we commit ourselves on the basis of simple tests, for we seem to be guided by the same biological wariness that animals exhibit before they gain confidence that they are safe in a new situation.

Should a field be regarded as a kind of matter? How do we define matter? As something in space and time that produces localised effects? For example, it denies space to other matter, it is a source of force producing effects distant from itself, it acts on light and other radiations. Need this action be uniform or continuous? And what are we to say about the existence of the matter used to test these effects? Is it uniform and continuous? At the microphysical level the existence of atoms and our need to rely on statistical interpretations shows us that uniformity and continuity are achieved by statistical averaging and smoothing, and that the important statistical idea on which we must rely in representing fields is the correlation of probabilities.

In modern physics we use the idea of virtual particles that are imagined to be created for the duration of the interaction process and then annihilated, energy being not necessarily conserved in the interim. Thus the matter of fields is created and annihilated. Whereas in ordinary matter the atoms exist as enduring entities under the scrutiny of the processes we command today to examine them, except in the rare instances when an atom is transformed by nuclear radiation, the atoms which we call elementary particles in the matter of fields exist in a different mode. We represent them and use our representation differently.

The modern concept of interacting fields, which is referred to in these pages as the New Aether, confronts us not merely with the philosophical question about existence, but also with processes that are quite foreign to our traditional metaphysical presumptions. These presumptions are most seriously challenged by the statistical aspects of the representation. Vacuum is not really empty—but we have not yet been moved to re-animate the contention of Descartes that Pascal demolished. Fluctuations in the electron-positron spinor field produce electromagnetic effects which in turn react on the former in networks of connexions. Since the elaborate theory of them leads to new and successful refined representation of the results of precise experiments over more than a decade, it has established itself as other great theoretical inventions did in the past.

The observed processes in nature on which was based the theory that has evolved into quantum electrodynamics are the Compton scattering of X- and γ-rays, the production of electron-positron pairs when γ-rays are 'absorbed' by the nuclear electric field of atoms, the inverse mutual annihilation of these pairs to form γ-rays, and the emission of bremsstrahlung from the collision of electrons with each other or with other charged particles. Each of these processes has been thoroughly studied in the laboratory and the theory of them has been well worked out. Why is it, then, that some of us should find it much more agreeable to accept the kinetic theory of gases in forming a physical idea of a gas than to accept the aether of quantum electrodynamics? Surely the answer must lie primarily in the simple nineteenth-century model of the molecules of a gas and the use of classical mechanics to treat collisions between them. Today, of course, the representation of a gas is quite different from this early model, because collisions must be treated quantum mechanically; we have to give up our classical picture of the collision of ordinary objects, treated statistically as if we could assign precise paths and speeds of motion to colliding pairs. But we have no doubt of the persistence of the colliding molecules even though we cannot localise them with classical precision in their motion. We are really not disposed physically to assume that under ordinary physical conditions molecules are destroyed* and created as we imagine elementary particles to be in the interaction of fields.

In quantum electrodynamics, however, we are compelled to represent the appearance and disappearance of electron-positron pairs, which seem real enough in experiments to study them—we can see their tracks in an expansion chamber or in a photographic plate, and are familiar with their many properties through a great number of varied experimental investigations. So we have learned to accept this departure from the ordinary notion of physical things. It seems much more likely that the source of our difficulty is the teaching of quantum mechanics about representing the motion of elementary particles. Since the interference of waves

* Not merely dissociated.

plays an essential role in computing the statistical results, and since no localisation is permitted between source and detector of the particles being represented, the entities we want to imagine in ordinary space seem to elude our comprehension. We represent distributions in space connecting the significant events but we must refrain from demanding the kind of picture which our ordinary experience disposes us to expect. We are unable to associate particle existences with events in the classical way. Since this is the way on which we rely in our use of ordinary language, we are confronted with a philosophical difficulty. It can be removed only by changing our metaphysical ground.

In the quantum field calculation of transition probabilities and other physical magnitudes that can be measured in an experiment,* as has been already remarked, the changes produced in the observing system are represented quantum-wise through the field functional that carries all the measurable facts of the physical system. These changes are separated from the intervening microphysical processes represented in the calculation—as if there were a self-contained world which we imagine (cf. Hertz), into which we can introduce probes to perturb it and from which we receive the quantised output of interaction with our instruments when these are present. Nevertheless, even when they are not, we have to imagine some physical effects occurring that alter the odds in the evolution of the continuing total physical system.

In fact, we think about physical systems in differing ways, depending on what we have in mind to do with them and to them, and depending on the physical situation we imagine them to encounter. Contrast, for instance, our views of an electron in the instances cited below:

(i) an electron in the K-shell of an atom of oxygen in a protein molecule in the absence of X-rays (i.e. under circumstances where we are ready to ignore the physical relevance of their action);

(ii) an electron in a hydrogen molecule;

(iii) an electron in a hydrogen atom with the utmost spectroscopic refinements in mind;

* This is a very vague expression, be it noted.

(iv)   an electron in a metal crystal;

(v)   an electron emitted by a radioactive nucleus in an expansion chamber;

(vi)   an electron in the Compton effect;

(vii)   an electron in cosmic rays;

(viii)   an electron in a high-energy accelerator.

In each of these circumstances the physicist conceives of the electron as existing, but in what sense? Is it the same in all of the foregoing instances?

Why should we be preoccupied about the existence of the entities of quantum field theory? It is fairly evident that most theoretical physicists speak of electrons, photons, nucleons, mesons and hyperons and so on through the pictorial realism of the particle physicist who detects them and measures their properties. Likewise they depend on the classical attitude to physical nature in using ordinary language; they take for granted that in this context philosophical or metaphysical questions can be ignored with impunity. Only when the particular application of theory demands it do they have recourse to the abstract representation, and even there the language of particle physics is employed as a convenient way of describing the effect of the mathematical theory.

Modern physics deals with atoms and atomic connexions. A theory of them must be a statistical one for an atom is the terminus of investigation by the routes that reveal it as an atom. The only structures in atomic phenomena that we can reveal are the statistical laws of families of atoms, finding out those forms relevant to the circumstances in which the atoms appear in the real world. We really investigate contexts for atomic transactions.

When we imagine the aether of modern physics we depend on the statistical evidence of the experiments to formulate the laws of the theory and invent the models which, while conforming to these laws, can be projected on the world we observe.

We are thus led to recognise that representing the world solely by means of the transitory processes and transient events treated by physical theories is unreasonable. We always have in mind some fixed classical aids in the process, for the microphysical interactions take place against a background that is treated in no more detail than is required for the purpose of the theory. The representing situation has much in common with that in the molecular theory of fluids. There the continuous stochastic processes treated by means of the Planck–Fokker equation are viewed against the background of a classical fluid motion. In microphysics we depend on the more or less steady environment—some sort of statistical average that suppresses the fluctuations we do not want to introduce. We require this to apply ideas of existence: it is needed to support our conception of reality at the microphysical level.

This need for support has its analogues in using language and explaining meaning, in finding a basis for the intellectual life of man and in sustaining the living aspects of the organic world. Anything resembling vitalism is out of place yet a vitalistic hypothesis seems to provide an escape from the complexity demanded of a representing process that attempts to put down everything that appears relevant; we invent variables to stand for unspecified particularity and then treat these variables as things. This will not do when we are dealing with reality and what exists: these words have to be anchored firmly in the world of naïve behaviour, in what is taken for granted not for intellectual discussion, but for acting in real life commitments.

In the discussion earlier in this chapter concerning the meaning of the word 'existence', it was emphasised that we should resist the temptation presented by the grammar of ordinary speech to think of this word as standing for a property of the object said to exist. We must now add to this admonition the following further one. 'Existence' always operates in connexion with the use of a particular representation. If we are determined to have a classical picture, as we must if we wish to think of the entities of particle physics in the same world with ourselves, let us accept the reasonable restriction based on experience that we cannot have as sharp a picture as we think we should by the unlimited refinement permitted by the classical method. We recognise that an experiment to study elementary particle processes depends on a limited beam of incoming radiation which is directed on the target. That is where the process we are studying takes place. It is when we start refining the representation to imagine the interaction of one incoming particle with a single nucleon in the target* that we have to depart from the classical representation. When this system is treated by quantum mechanics, the waves seem to fill the space of the relative spatial coordinates of the interacting particles, and accordingly we seem to have destroyed the fairly good localisation of the target nucleon and to have spread the incoming particle over the whole of space. Much of this appearance of unreality is not to be taken literally, in spite of the tendency on the part of some theorists to write as if we should. The physical fact is that the interaction is localised within a volume of atomic or nuclear dimensions.† The infinite distance of the mathematician is merely one great compared with this. Our classical idea here is not nearly so inappropriate as one might be tempted to believe from reading how we must give it up in atomic and nuclear mechanics. The importance of seeing the relevance of classical forms of speaking about phenomena attaches to helping us to overcome the philosophical doubt that invades our minds about the existence of the entities named in microphysical representation.

We may approach this matter in another way to expose how we are lured into philosophic perplexity. Let us consider the conventional methods of presenting a new theory in physics. It is taken for granted that the reader of the scientific paper or the audience in the lecture has a good idea of 'the problem' which the theoretical invention is, if not to solve, at least to contribute to solving. The essential point of the communication is to show a new way of doing things. An effective invention attracts interest and very soon it is being elaborated in the minds of other men. The influence of the idea tends to grow. Sooner or later it collides with other spheres of influence of basic ideas, that have a different history. The conflicting ideas have to live together until some measure of sorting out is achieved and men learn the proper scope of their effective-

* We ignore here diffraction of the incoming radiation by a crystalline target.
† In the coherent scattering that produces diffraction by a crystal lattice the localisation is of course extended to the volume of the microcrystalline diffracting elements.

ness. In mathematical theory, however, we must have a consistent system. Accordingly, when new ideas have to be accommodated, a substantial rebuilding of the mathematical structure may be required. That is, the usual thing is to replace the old model with a new one. In mathematical writing this may seem clear enough, but physics is more than a mathematical theory. The self-contained system of the formalist has to be supplemented in many different ways to use it, indeed, the complexity of these elaborations soon outruns our ability to keep formal account of all of them. Thus our attention is concentrated on what we can manage, the symbols of the formalism, and very soon we may succumb to regarding them as the reality. We mislead ourselves, however, if we imagine that there is a regular procedure in innovation, for its ways are as complex as the varieties of personality and behaviour of the men who are engaged with them. It is when we try to present a certified picture of what the experts have 'established' that the lures of formalistic uniformity inveigle us to treat a consensus of informed opinion as something fixed and immutable, and a permanent memorial to the originators of new views.

Of the idea of an aether, even of our readiness to use the word, enough has been exposed in these pages to recall what should be evident to everyone familiar with the history of physics, indeed of all science. Authority, convention and dogmatic formal exposition all have their place in the communications of men about science, but the fields on which they operate are not fixed except for a short time. Over the years, knowledge grows, attitudes change, and in the process the theoretical fabric of physics is renewed in ways that do not easily lend themselves to simple analysis and classification. Examining these matters is, in a sense, a metaphysics, a study about physics, concerned with how we regard and invent and choose formal inventions for physical theory. Its purpose should not be to generate formalism, but to encourage a sane resistance to misplaced formality. Such considerations impinge on our epistemological prejudices about how we learn, how we think, and how language operates. The physical scientist need not be an expert in these matters in order to benefit from the scepticism which understanding more about ourselves may be expected to induce towards scholastic formality, as distinct from proper effective formalised techniques.

To comprehend the New Aether, and adapt ourselves to the changed world that the ideas on which it is based have made for us, we need confidence that we can commit ourselves to it. Doing so involves us in much the same kind of experience that men in the world of the fifteenth and sixteenth centuries had to go through to accept the discoveries of the great navigators. Our maps are not the same as their maps, but no sensible man doubts that they spoke and we speak of the same globe. Modern men have observed evidences of changes that have taken place on the Earth in the intervening centuries, and discovered many details unknown to the Spanish and Portuguese explorers. 'Mais le monde existe depuis longtemps.' Whatever uncertainty we may entertain about the life of physical theory we have no choice but to accept the best we know as the effectual basis for improving our understanding of our environment. And so it is with the New Aether. The empty space in our simple geometrical representation of a vacuum has gone and made way for the random microphysical activity which our statistical view of atomic nature demands.

Thus our picture of nature reflects our experience in using sensitive physical instruments. As we make them more sensitive, sooner or later we encounter the phenomena associated with noise, due either to complicated physical causes in the environment that only weakly affect the instrument, or to the random heat motion in the matter of which the instrument is made. Noise sets a limit to the refinement of measurement and therefore bounds the field of application of the theories to which the results of measurement are relevant. Nevertheless, we have physical theories of the statistical characteristics of noise, based on imagined microphysical processes, and employing the successful methods of quantum mechanics to compute not only the statistical averages, but also the asymptotic forms of the connexions between them.

In quantum electrodynamics under the influence of the perturbation method one proceeds from the 'free fields', in which the interactions between them are ignored, by introducing these inter-actions in progressively higher orders. Their effect is to 'clothe' the original 'bare particles' with the dress of the properties they acquire as the result of their interactions with the other fields. Thus the coupling of the electron-positron field with the photon field leads, for example, to a value of the magnetic moment of the electron slightly different from that presented in Dirac's rela-tivistic theory of the electron. The computed value of this quantity agrees very well with the corresponding measurement.

When one thinks of an electron as an atomic connexion between events the representations which induce us to speak of the bare particles of the free fields and the dressed particles of the inter-acting fields, seem somewhat out of place. Just as the quantum mechanical treatment of motion appears to represent something going on between the emission and the capture of the particle, whereas we should understand that the purpose of the continuous representation is to establish the statistical distribution over the appropriate family of atomic existences, so we should recognise that the interaction of fields establishes connexions between the system of ingoing particles and that of the outgoing ones. These connexions may be represented as intervening in space and time, as they are in our treatment of the scattering of particles, or of the nuclear reactions caused by them. In doing this, we are com-mitting ourselves to the hypothesis of a physical process in space-time, which is how we regard the matter classically. The incoming particle hits the target and something happens. Since a localisation occurs in this event we may think of an atomic existence bounded by emission from the source and hitting the target followed by emission from the target and detection at a distance. That is, we have a different kind of atomic connexion, namely, the stochastic connexion of atomic existences in a scattering collision or reaction process. In so far as the collision event is itself not specified on the spatio-temporal scale characteristic of interacting physical entities and their fields of force, but only with the degree of discrimination adequate for the distinction between events in the laboratory, this classically defined event is not relevant to limiting the atomic existence in the sense of destroying the coherence between the various alternatives which our continuous representation of the physical system of interacting particles presents to us. In the mathematical theory, we ignore the limitation, so the scattering or other phenomenon is presented as if the atomic connexions between source and detection events remained intact. The processes we imagine with the aid of physical models, subject to the formalism of quantum field theory, serve only to connect atomic existences by statistical laws governing their distribution and frequency of appearance that can be compared with experiment. This decides how appropriate the theoretical method is. Whether we are disposed to go farther than this is determined by the way in which the successes of the theory can be used as the basis for experimental invention to push investigation farther.

The condition logically imposed on atomic existence is clear. The existence ties together events in space-time under conditions that make the interference of waves appropriate, because in the corresponding classical representation, we have to go within the limits prescribed by the uncertainty principle. But this does not prevent our using atomic existences to derive a picture or other physical representation of the physical objects we investigate by studying the statistical effects of their interaction with the atomic existences in question. That the statistical description is outside the particular atomic existence does not affect the validity of the argument. Let us recall that we make a diffraction grating using devices and processes that are not limited by the degree of precision permissible if we depended only on the waves of the wavelength which we propose to use in producing diffraction by means of the grating. Yet the diffraction pattern with the waves informs us about important aspects of the structure of the grating. Thus light, X-rays, electrons and neutrons have served to reveal by interference elements of the structure of matter that we have been unable to show as effectively by any other means.

We are concerned in the present discussion with how we speak of these matters, because this is how we draw into our thinking metaphysical attitudes that do not help us in doing physics. Our fundamental difficulty lies in attaching an atomic existence to the continuous representation of events in space-time and the continuous field representations that we have learned to associate with it. We achieve continuous representation through the statistical laws that govern the connexion between the families of atomic existences and the world of experience that we represent classically.

The important purpose served by our philosophical activity here is to make clear that the modern physical pictures have the force of classical pictures because we use them as classical. For example, consider a graph of the electromagnetic structure of a nucleon.* This represents the density of scattering power for the billion-volt electron waves used to explore the proton and neutron. The spatial distance, the abscissa in the graph, is the relative distance of nucleon and electron. But this classical picture is essentially a statistical picture. The object is defined by its effect on the beam of electrons, not on a single electron. To what extent does this correspond with our view of an ordinary object which, we should remind ourselves, is a symbol for the unspecified possibilities of experience in which 'knowledge about the object' plays a part in our thinking and action?

The 'statistical picture' above means a picture with only a statistical application. It is quite clear that in physical literature today, such pictures form the basis not only of progress in the subject (in spite of Dirac's famous passage outlawing them), but also in presenting physics to the non-specialist in theoretical physics. Whatever qualifications and reservations the professional theorist may have in regard to the validity of these pictures, these reservations are brushed aside in the context where the pictures are of real use, viz. in aiding appreciation of physics and getting a view of the physical world as revealed by modern physics.

In discussing the processes by which elementary particles interact, physicists use the diagrams that Feynman introduced in quantum electrodynamics. These diagrams present possible schemes of atomic connexions. Indeed, in an important respect

* R. Hofstadter and R. Herman, 'The Electric and Magnetic Structure of the Proton and Neutron', *Physical Review Letters*, **6** 293–6 (1961).

we learn from the use of these diagrams what constitutes an atomic connexion. The diagram presents the order in time of the events connected by electrons, photons, mesons, nucleons, and so on. It is not concerned with representation in space, except to indicate incoming and outgoing particles. The mathematical theory, however, deals with events represented in continua for the quantum theoretical calculation of the probabilities of experimentally distinguishable results of interaction. By looking away from the details of the mathematical calculation we avoid trying to think of the atomic connexions as if they were particles travelling through space. We are dealing with imagined connexions between events imagined in the continuum; but the lines that denote ingoing and outgoing particles represent connexions to the real world of the experimental physicist with his apparatus to produce the phenomena and his instruments to investigate and measure. The aether is an imagined world of atomic connexions between the real things and processes that the physicist controls and observes.

CHAPTER X

ACCOUNTING OPERATORS

Among the forms that characterise physical representation by means of quantum mechanics are logical symbols that name possibilities in the spaces of physical variables that can be measured. For example, $\delta(E - E_n)$ denotes the logical possibility that E takes the particular value E_n. A statistical distribution over a particular space is represented by means of a linear combination of these symbols. The coefficients attached to the symbols in this combination are in general complex numbers. The squares of their moduli are the probabilities that the corresponding possibilities are realised, and the phase distribution exhibited by the set of complex coefficients is relevant to the superposition of one distribution on another. We have already discussed how the phase takes account of the atomicity of action in a mathematical system that depends on continuity to support the analysis. Quantum mechanics provides effective apparatus for ensuring that without transgressing the limitations imposed by atomicity, the discrete and continuous can be treated in association with continuous representation. It is quite natural, therefore, that we may be tempted to identify the continuity in analysis with the continuity we associate with our usual representation of physical nature in space and time. The continuity of the latter representation applied to atomic phenomena is effective only through statistical interpretation. It is when we overlook the statistical implications of the quantum mechanical method that we are confused and misled by their classical counterparts.

We may imagine physical phenomena as if there were a wave process—like a classical wave motion in fact. But if we encounter metaphysical difficulty about this we must go back and extricate ourselves from the implications of this too naïve attitude to the quite sophisticated mathematical invention.

The effect of these considerations is to exhibit the logical implications of accommodating the atomicity of the variable action in the mathematical system that depends on continuity to support the analysis.

Quantum mechanics has been developed to a very great extent by elaborating the mathematical possibilities opened up by its methods. Merely to make the point about the relation of the phase of a wave function to equivalent lattices of possible values of the argument serves only one purpose, namely that we should gain some insight into the interference of probabilities not as a transcendental aspect of physical nature we have just to accept, but as a necessary concomitant of representing atomicity by means of continua. The waves are our invention. Heisenberg's principle prevents us from achieving the kind of representation that would reveal the wave motion in space and time as the variation of a measurable function or system of functions, for the quantum theory of fields restricts our simultaneous specification of amplitude and phase.

The classical view of an actual electromagnetic field associates with it energy and momentum stored in space and transmitted through it. So the field itself is a dynamical system. Since it is a continuum, the number of dynamical variables required to represent it is infinite. This conception is applied not only to the electromagnetic field; it is applied to fields with different structure and particular roles in the description of physical nature. We start from an appropriate model in which the dynamical structure of a continuous field is exhibited by presenting the spinor, vector, tensor or other system of functions of space and time as variables for specifying the field. With an appropriate form of the Lagrangian density, we derive from the stationary property of the action of the field, the differential equations satisfied by the field variables. Then we proceed to extend the use of the formal methods of quantum mechanics to this system. The forms adopted to describe the field are chosen compatible with the requirements of the special theory of relativity. The formal process for transforming the representation of the field so as to introduce the discreteness

169

characteristic of atomicity was called 'second quantisation'. In analogy with the formal procedures for passing from the dynamical variables in the classical sense (c-numbers) to the corresponding variables in the quantum sense (q-numbers) the field variables are regarded as operators on the field functional. This functional is imagined to play a role analogous to that performed by the wave function of a dynamical system whose configurational coordinates are finite in number. However, the analogy is by no means complete. Whereas in the quantum theory of atomic spectra we have to deal with explicit forms for the dependence of wave functions on spatial coordinates and for some purposes also on the time, in dealing with the quantum theory of fields we are rarely involved with the corresponding explicit expressions for the functional in terms of the field variables. The field functional is merely an accounting symbol presenting a table of occupation numbers in the particle states which are named by momentum vectors of the particles.*

Once an appropriate system of field variables has been chosen— in the electromagnetic field this system is the electromagnetic potential 4-vector—the variables are represented formally by means of Fourier analysis in space and time when a plane wave representation is chosen. Other systems of characteristic waves may be used and the wave functions suitably normalised. Thus each field variable is represented as linearly dependent on the normalised characteristic forms which are functions of space and time. The coefficients in this linear expansion then become the objects of interest as a new set of dynamical variables.† They are treated as operators subject to commutation rules just as the momenta and configurational coordinates are in non-relativistic quantum theory.

In the plane wave expansion are found pairs of characteristic wave functions which differ only in the sign of the exponent of

* Since the mass of the free particles of the field is supposed known, this specification suffices to determine the corresponding energy of the free particle.

† Cf. algebraic methods of treating the dynamics of a vibrating system by using normal coordinates.

the function e^{ikx} that represents the wave.* The corresponding coefficients in the expansion are denoted by b_λ and b_λ^+, where λ stands for the set of numbers specifying the momentum state. The operators are defined by

$$c_\lambda = \sqrt{\frac{2\omega_\lambda}{\hbar}}\, b_\lambda, \quad c_\lambda^+ = \sqrt{\frac{2\omega_\lambda}{\hbar}}\, b_\lambda^+,$$

where $\hbar\omega_\lambda$ is the energy corresponding to the momentum state λ, and the amplitude of each characteristic wave is normalised over a large but finite volume V. It is mathematically convenient to define the operators in this way because the commutation rules then take the simple forms associated with the two types of field, one (boson) symmetrical with respect to the operation of interchanging particle labels, the other (fermion) anti-symmetrical with respect to that operation.

For a boson field the commutation rule is

$$c_\lambda c_\lambda^+ - c_\lambda^+ c_\lambda = 1.$$

All other pairs of c's and c^+'s commute. If $\Phi(N_1, N_2, ..., N_\lambda, ...)$ denotes the field functional, N_λ being the number of particles in the single particle state λ,

$$c_\lambda c_\lambda^+ \Phi(N_1, ..., N_\lambda, ...) = (N_\lambda + 1)\Phi(N_1, ..., N_\lambda, ...),$$
$$c_\lambda^+ c_\lambda \Phi(N_1, ..., N_\lambda, ...) = N_\lambda \Phi(N_1, ..., N_\lambda, ...)$$

and it is easily shown† that

$$c_\lambda^+ \Phi(N_1, ..., N_\lambda, ...) = \sqrt{(N_\lambda + 1)}\,\Phi(N_1, ..., N_\lambda + 1, ...).$$

That is, acting on the functional representing a state of the field in which the particle state λ is occupied by N_λ particles, the operator c_λ^+ alters the occupation number of the state by adding one particle and by changing the amplitude of the field functional. Thus we may think of c_λ^+ as a particle creation operator, and correspondingly c_λ as an annihilation operator on the table of occupation numbers presented by the field functional. Clearly the

* k and x are 4-vectors, kx denotes their scalar product; k is $(\hbar)^{-1}$ times the energy-momentum vector, x is the vector displacement in space-time; k_0 and x_0 denote the time-like components of the vectors.

† Akhiezer and Berestetsky, *Quantum Electrodynamics*, p. 144.

energy associated with the state having the greater number of particles exceeds that for the smaller number by $\hbar\omega_\lambda$, where ω_λ is the angular frequency of 'time dependence' in the harmonic wave λ. It is for this reason that the change in amplitude of the field functional is induced by c_λ^+.*

Even if in our representation our creation operators imported every physical variable that we imagine relevant to the existence of the entity that we represent as having been created, the operator does not create anything in reality. Acting in the representation it stands for the appearance of a new form in a discontinuous change. There is imagined no process of creation that can be itemised like the organisation of form by genetic replication. In the process of β-emission from a radioactive nucleus, the electron suddenly appears. In accordance with our atomic hypothesis, we do not attempt to interpolate a physical process between the non-existence of the electron and the electron leaving the nucleus. Such a process could only be supported by something which undergoes change in the birth of the electron. Mechanically this appears to be out of the question because of the uncertainty principle. Creation, just as motion, is an atomic connexion; it cannot be dissected. Nevertheless, if it were discovered experimentally that the statistics of creation could be affected by means of an agent affording the means of sharp timing, we would examine the possibility of representing this action in the space-time continuum as we treat the interaction of particles in quantum field theory.

Let us now return to the commutation rules for the creation and annihilation operators and consider the fermion field. We have to introduce not merely the operators associated with the spinor

* We may note in passing that since we represent a field by a linear combination of waves, and since a plane wave can be represented as the result of superposing two standing wave systems of equal amplitude in phase quadrature both in space and time, the field may be thought of as a set of standing waves each of which is in effect a harmonic oscillator. The energy of the field is the sum of the energies of the oscillators $\Sigma_\lambda \, N_\lambda \hbar\omega_\lambda$, provided that we define the zero of energy so as to remove the zero-point energy $\frac{1}{2}\hbar\omega_\lambda$ of each oscillator. Unless the number of occupied possible modes λ is limited, the energy of the field is infinite.

waves that represent the particle states, but also the operators associated with the spinor waves that represent the anti-particles, for instance, in the best known fermion field, the electrons are the particles and positrons are the anti-particles. We denote the former by a_r^+, a_r and the latter by $b_{r'}^+$, $b_{r'}$, respectively. On account of the Pauli Principle governing the occupation of the particle and anti-particle fermion states, only two possibilities have to be considered for each state—either the state corresponds to the presence of one particle, or to the presence of no particle with the dynamical momentum and spin appropriate to the state in question, which we name by the subscript r.

Except the particle creation and annihilation operators with the same subscript, and the anti-particle creation and annihilation operators with the same subscript, all of these operators anti-commute; for example,

$$a_r a_s + a_s a_r = 0, \quad a_r^+ b_s + b_s a_r^+ = 0;$$

but

$$a_r a_r^+ + a_r^+ a_r = 1 \quad \text{and} \quad b_r b_r^+ + b_r^+ b_r = 1.$$

From these rules it is deduced that the compound operator $a_r^+ a_r$ can take only the values 1 or 0 corresponding to the occupation or not of the rth particle state, and similarly for $b_{r'}^+ b_{r'}$ for the r'th anti-particle state. We may represent a_r^+ and a_r as matrices operating on the 2-component vector whose components stand for the logical possibilities: ψ_0, state unoccupied, and ψ_1, one particle present. We omit the subscript r since we are thinking only of one state. Since

$$a^+ \psi_1 = 0, \qquad a \psi_1 = \psi_0,$$
$$a^+ \psi_0 = \psi_1 \quad \text{and} \quad a \psi_0 = 0,$$

the appropriate matrices are

$$a^+ = \begin{pmatrix} 0 & 1 \\ 0 & 0 \end{pmatrix} \quad \text{and} \quad a = \begin{pmatrix} 0 & 0 \\ 1 & 0 \end{pmatrix}.^*$$

The operators on the anti-particle states b^+, b are similarly repre-

* Note that $a^+ a^+ = aa = 0$.

sented. We may represent these operators as linear combinations of the Pauli matrices

$$\sigma_1 = \begin{pmatrix} 0 & 1 \\ 1 & 0 \end{pmatrix}, \quad \sigma_2 = \begin{pmatrix} 0 & -i \\ i & 0 \end{pmatrix}, \quad \sigma_3 = \begin{pmatrix} 1 & 0 \\ 0 & -1 \end{pmatrix}$$

as follows:

$$\sigma_1 + i\sigma_2 = 2a^+, \quad \sigma_1 = a^+ + a,$$
$$\sigma_1 - i\sigma_2 = 2a, \quad i\sigma_2 = a^+ - a,$$
$$a^+a + aa^+ = 1, \quad a^+a - aa^+ = \sigma_3.$$

The purpose of exhibiting these relations here is twofold; first to bring to light that certain operators change the phase of the wave function,* and secondly, to indicate that the mathematical equations connecting the operators are indeed logical connexions. Thus we may read for σ_1, the logical addition of a^+ and a, that is σ_1 stands for creation or annihilation. Which of these operations can happen is determined by the state on which they operate. If the state is occupied, only annihilation is possible; if not, only creation is possible. The operator σ_2 stands for annihilation or creation, the resulting state being advanced in phase by $\frac{1}{2}\pi$ for the former, and retarded by $\frac{1}{2}\pi$ for the latter. The operator σ_3 stands for annihilation followed by creation, or creation followed by annihilation with phase reversal. Finally, since one of a^+a and aa^+ must destroy the state on which it operates, while the other leaves it unchanged, their logical sum must leave the state unchanged.

The square of each of the operators σ is unity. This may be exhibited as a tautology as follows:

$$\sigma_1^2 = (a^+ + a)(a^+ + a) = a^+a^+ + a^+a + aa^+ + aa = 1.$$

The first and fourth terms on the right vanish as we have already noted. The second and third make up the combination we have discussed.†

We have introduced the Pauli matrices in this context because of the logical simplicity of the situation. Their appearance in the representation of rotation is much more subtle, and their use in

---

* Cf. boson case where change of amplitude was noted.

† That the eigenvalues of σ_3 are 1 and -1 is immediately obvious in terms of the equation connecting that operator with a^+ and a.

connexion with isobaric spin space is a still different application. In each of these, however, the elements of the matrices connect pairs of logical possibilities (not necessarily different). If the algebraic combinations are interpreted as logical formulas, we see how the algebraic rules allow only those operations that are consistent with the Pauli Principle, and permit interpreting the operator $u^! u$ as the operator representation of the number N_r of particles in the state r. N_r may take only the value 0 or 1. Likewise $b_r^+ b_{r'}$ is the operator representation of the number of anti-particles in the state r'. Since particle and anti-particle carry equal and opposite electric charges, the charge in the field is represented by the sum over r and r' of $a_r^+ a_r - b_r^+ b_{r'}$ multiplied by the charge of one particle. The energy of the field is given by the sum of $a_r^+ a_r + b_r^+ b_{r'}$ multiplied by $\hbar \omega_r$.

If one examines in further detail this apparatus for representing fields corresponding to free particles one finds mathematical artifices on every hand: their effect is to guarantee that the energy of the field is tied into the bundles we associate with the particles. The plane waves are normalised so that the energy of the individual characteristic wave is $\hbar \omega_r$. But the normalisation depends on the choice of the operators to give the commutation relations their particular form, so the quantum expression for the energy is introduced mathematically.* The mathematical theory does not explain how the classical quadratic dependence of the energy quantum on frequency should be transformed into the quantum linear dependence, it merely incorporates the quantum rule of Planck into the algorithm. This remark is intended to draw attention to the artifices of mathematical invention by which the accounts are kept in accord with experiment. The theory does not throw any more light on this matter than is given in the bare hypothesis of Planck. It has evolved the form that leads to agreement with experiment. For this reason one begins to wonder just how relevant are the classical fields in the context of quantum physics. Perhaps the theoretical

---

* As indeed Bogoliubov and Shirkov state explicitly, *Introduction to the Theory of Quantised Fields*, p. 115, trans. by G. M. Volkoff (New York: Interscience Publishers Inc., 1959).

difficulties that appear to Landau to require us to give up the apparatus of field functions really stem from the shot-gun marriage of the classical field with the mathematical formalism of quantisation. Whereas half a century ago some physicists attempted to impose the quantum law $E = hf$ by means of models, today we do logically the same kind of thing in a much more sophisticated way, but in a formal scheme that steps out of the classical modes of physical thinking. The point of the model was to 'explain' the implications of the rule and make it palatable. This is no longer necessary because the formal compulsion has been accepted.

Since the creation and annihilation operators are subject to commutation rules that can be interpreted logically, we have to examine how they are combined to produce the symbol that corresponds to the physical variable in the classical field, so as to see what the formalism is doing for us. The essential point may be illustrated by showing the expression for the field function in a scalar boson field.* The positive frequency part of the wave function is

$$\phi^+(x) = \frac{1}{(2\pi)^{\frac{3}{2}}} \int \frac{d\mathbf{k}}{\sqrt{2\omega_\lambda}} \, e^{ikx} \, \phi^+(\mathbf{k}),$$

where $\phi^+(\mathbf{k})$ is the expression equivalent to c_λ^+. From the commutation relations governing the c-operators we can pass to those governing $\phi^+(x)$ and $\phi^-(x)$, the positive and negative frequency parts of $\phi(x)$, and thence to the commutation relations for the complete field functions, namely

$$\phi(x) \, \phi(y) - \phi(y) \, \phi(x) = i^{-1}D(x-y),$$

where

$$D(x-y) = \frac{i}{(2\pi)^3 i} \int e^{ikx} \, \delta(k^2 - m^2) \, \epsilon(k^0) \, dk$$

and

$$\epsilon(k^0) = 1 \quad \text{for} \quad k^0 > 0, \ = -1 \quad \text{for} \quad k^0 < 0.$$

* See Bogoliubov and Shirkov, *Introduction to the Theory of Quantised Fields*, pp. 114 *et seq.*, trans. by G. M. Volkoff (New York: Interscience Publishers Inc., 1959).

The symbol D is a singular improper function; it is the solution of the wave equation

$$\nabla^2 D(x) - \frac{1}{c^2} \frac{\partial^2 D(x)}{\partial t^2} - \frac{m^2 c^2}{\hbar^2} D(x) = 0$$

under the boundary conditions

$$D(x) = 0, \quad \frac{\partial D(x)}{\partial t} = \delta(x) \quad \text{at} \quad x^0 = t = 0.$$

In this way we establish a formal condition on the logical operators ϕ which is interpreted by expressing ϕ explicitly in terms of the creation and the annihilation operators. We may think of $\phi^+(x)$ as the operator that switches on the distribution of probability amplitude given by the form of $\phi(x)$ and $\phi^-(x)$ as the operator that switches off the same form. Thus we may speak of the creation of a particle in a particular state of motion, and the annihilation of the particle. In the theory the operator $\phi^-(y)\phi^+(x)$ creates the particle at the event denoted by x and in the state ϕ and annihilates it at the event y. So it corresponds to the propagation from x to y. Physically we should require that the event x should antecede y because creation must precede annihilation. Accordingly, we should arrange the mathematical form so as to take this into account. For this reason chronologically ordered products of operators are introduced and related by analysis to 'causal Green's functions'. These are solutions of the inhomogeneous wave equation

$$\left(\nabla^2 - \frac{1}{c^2} \frac{\partial^2}{\partial t^2}\right) D^c(x) - \frac{m^2 c^2}{\hbar^2} D^c(x) = -\delta(x)$$

which necessarily vanish when the time component of the argument of D^c is negative. In this way we reach the remarkable result that the ordered product of the two operators is mathematically equivalent to a singular function of the relative coordinates of the events at which the operators are applied. Since this function is a mere multiplier of the state vector of the field, it alters the latter only in amplitude and phase, but only provided that the events connected by means of it are properly ordered in time; so the singular function is an analytical device that automatically secures this time ordering. The probability of the process represented by the chronologically ordered product connecting two regions of space-time is calculated by summing over them and taking the squared modulus of the sum. The algorithm for more elaborate products and sums of products has, of course, been developed.

To deal with fields of other types the analysis is elaborated to cope with the fact that more than one function of space and time is needed to specify the corresponding classical field. The idea of polarisation is introduced to distinguish the enlarged system of possibilities just as is required to represent a classical vector field in terms of the system of vector characteristic waves. In principle, however, the notation is expanded naturally to supply the additional individual labels. Correspondingly, the commutation rules appear more complicated, but this is a complication required only by the enlarged notation. The field equations are different and the Green's functions correspondingly altered. But, in principle, the operators corresponding to the classical field functions are linear combinations of the creation and annihilation operators for the different momentum and polarisation states. The characteristic waves are normalised in conformity with the requirements of the quantum rules for energy and momentum.

To proceed farther in discussing these matters involves us in greater detail with quantum field theory. For our purpose it suffices to say that, in representing interacting fields, it is necessary to assume a particular form for the interaction term in the Lagrangian density for the coupled fields and to pass from that to the transformation that connects the initial state of the incoming particles with that of the products of interaction. The transformation is represented by combinations of operators appropriate to the fields and to certain rules for combining field components. Methods exist for calculating any magnitude that is normally measured. As to the mathematical difficulties that the theoretical calculations encounter the reader is referred to the works already cited.

The forms to which we have just alluded are chosen compatible with the theory of relativity and to conform to the restrictions imposed by experimental knowledge about the interactions. Algebraic methods of pervasive, compressive power dominate analysis and give the theory a very abstract appearance, so that the mathematical theory seems to apply to a strange world. When one approaches it in a naïve way, treating the signs as intended to represent things or connexions in the world, it does indeed seem strange. But this strangeness disappears when the contractions of notation are recognised for what they are, and the mathematical details are exposed, learned, and understood in the way in which mathematical methods are usually effectively mastered. One has to remember that once learned a mathematical system forces itself on our thinking, because of the economy of expression its notation gives us. Details that must be exposed to remove philosophical difficulties often do not come to light at all in the calculations using the algorithm. Eventually, physical phenomena come to be spoken of through the mathematical theory; in the process the abstractions of theoretical connexion acquire concrete associations when they are applied physically to the substance of the real world. There is really no mystery about these things when we keep our metaphysical balance and remind ourselves that theoretical invention is the work of men who not only have had a long training to acquire the skills on which they depend, but who, except for rare exceptions, are not interested in explaining what they are doing unless within the narrow bounds of formal mathematical instruction in physical theory.

Our concern here is to disintegrate the concept that there is something occult or magic about the success of quantum mechanics. As if only mathematics could cope with the mystery of nature! There is enough evidence that quantum mechanics has grown up, as have other mathematical theories, supported mainly by inventions for manipulating the mathematical signs. The relevance of the logic of the system to physics has been only dimly illuminated. The physical concepts that support the application of quantum mechanics are essentially classical. To say that the physical variable has to be treated as an operator and that its measured magnitude must be an eigenvalue of this operator seems to suggest that in the laboratory we must remain ignorant of the connexions between events that the mathematical theory somehow or other manages to encompass. As a matter of historical fact, the inventors of quantum mechanics were led to their inventions through classical physics and a thorough knowledge of the experimental facts about atomic spectroscopy. They created a very satisfactory accounting system that has evolved in many unexpected ways. But the metaphysical questions associated with quantum mechanics have not been really answered. If they had, then we should understand why it is that quantum mechanics works in the way it does.

The mathematical situation is not simple, it requires us to understand how the algebra of operators is connected with the functional calculus,* and to see how the analytical difficulties of improper functions can be overcome completely satisfactorily in the theory of distributions. Since theoretical physics attracts the interest of mathematicians the subject will never lack formal inventions. To incorporate them in the thinking of physicists it is necessary to explain what each invention makes possible for us in changing our method of thinking about physics, and to discover the simplest way to give it effect.

That the wave function in a field is a logical symbol would be more clearly evident if we had to deal with possibilities on lattices only. Then the independent possibilities would be named by sets of integers and the symbol would not be presented as a function of space and time, for instance, and then it would not resemble the mathematical representation of a classical physical variable whose magnitude can be measured. From these basic possibilities we can generate others represented by linear combinations of them, that are logical sums in which addition stands for 'or', consequently the combinations are to be interpreted statistically.

Having set up a system of possibilities on one lattice we have to consider the possibilities with respect to another. If we can set up a one-one correspondence between the lattices point by point then

* See H. Feshbach and P. M. Morse, *Methods of Theoretical Physics*, chs. 7 and 8 (New York: McGraw-Hill Book Co., 1953).

clearly we are in no logical difficulty; we deal with the same system of possibilities in the two representations.* If, however, the possibilities represented by the points of one lattice (A) do not correspond with the points of the other lattice (B), but correspond to linear combinations of the logical symbols for the possibilities of B, the two systems are not compatible except in the following sense: a possibility of A is only a statistical possibility with respect to B and conversely.

Some of the philosophical questions associated with quantum mechanics have the disconcerting force of the questions about the logic of 'number' discussed by Frege, Russell, Wittgenstein and others. We should be warned therefore by what has happened in mathematics. Words with varied application like 'number' or 'space' may lead us into logical confusion. Such a word in the core of quantum mechanics is 'state'. So long as we have to deal with a fixed number of particles we can enumerate the dynamical variables of the system and write down the characteristic forms of the eigenfunctions. A particular state is specified by giving its scalar products with a complete set of eigenfunctions. This means giving the laws for generating these scalar products because their number is infinite. When we come to quantum field theory we speak of the state amplitude or state vector. The number of particles is now indefinite, because particles are created and annihilated in physical processes. All that the state vector does is to act as a sort of store in which is kept the score of the changes produced by the operations that are made on its contents. In analogy with using a computing machine we might look on the state vector initially as an input operation on the store, the operators are the computing instructions which indeed they are, and the adjoint state vector is the output instruction, so that $\langle \Phi_2 |$ operation $| \Phi_1 \rangle$ stands for the result from the whole process. An eigenstate of a particular operator is a store content that is unaltered by the operator and on output yields a particular number, the eigenvalue.

The advantage of this model for thinking about the subject is

* Cf. commuting variables with common eigenvectors.

that it removes the temptation to regard the mathematics as a picture of the physical system being treated. It is an accounting system. Some mathematicians seem to say that we cannot have anything but the accounting system. Nevertheless, we know very well that an accounting system may be given an effective life when it is used in human affairs. It is the use of the accounting system of modern theoretical physics that confers on it physical significance. The reality represented with its aid is not to be found as a substratum governed by the mathematical formality: for the essential real supports are found in the real world of men and machines and physical instruments and phenomena produced and observed by means of them in the lives of the men who use them. These provide the proper contexts in which the words 'real', 'exist' and 'substance' are used. Instead of asserting that the mathematical theory represents reality, we do better to say that we use the theory in managing our experience in the real world with more or less success. For apart from its use, theory is not brought into contact with physical reality.

Now let us look at the state called the vacuum state. For accounting convenience it is attractive to define it as subject to $P|o\rangle = o$, where P is the momentum operator, and it seems reasonable enough that we should not expect to measure the momentum of vacuum. It is attractive also because the equation implies that spatial displacement of the system (vacuum, or state) does not alter the state.* By means of this definition we eliminate from the expression for the energy, the zero-point energy for the vacuum state, but are we really satisfied that this is correct physically? Can we ever have a space free from physical processes? Of course not. We must introduce noise to stand for our ignorance, so the idealist concept of a vacuum state must be recognised for what it is—an aid to mathematical theory. It does not stand for a physical vacuum, for we believe that quantum fluctuations are occurring all the time—virtual photons being created in the electro-

---

* Though how one can displace vacuum physically is a conundrum. It would make sense to say that the system looks the same whatever the location of the coordinate axes—which, of course, says very little.

magnetic field and virtual pairs in the electron-positron field. This is our physical view of the matter. When a real electron is physically introduced, it is accompanied in its motion by the virtual 'particle-wave' cloud induced by its presence; these virtual entities by their presence contribute to the properties of the electron. We ought to return to Hertz to correct our thinking on this essentially metaphysical matter. His expression was 'We have to imagine'. It seems we have to return to this simple confession and recognise that theorems of impotence which serve as mathematical axioms really do little for us in the long run. It is the application of their logic to our thinking about particular systems imagined or brought to light by technical ingenuity that advances science.

Our representation of microphysical phenomena may be regarded somewhat as follows. On the one hand, we depend on elaborate physical machines by which the phenomena are produced, isolated for study and investigated, using techniques and inventing new ones for detecting physical effects in events, and recording and analysing them statistically in the categories chosen by the physicist. On the other hand, the natural processes imagined to produce the effects sampled by the experimenter are conceived merely as formal supports for the individual atomic transitions (including motion) by which radiation is emitted, absorbed, propagated and so on. The imagined processes serve also to explain the properties of atoms, molecules, nuclei, and other entities with which radiation can interact.

Whereas according to the classical tradition every natural process is regarded as subject to continuous laws of connexion such as we use successfully to deal with our mechanical and other inventions, the continuous connexions for microphysical representation belong merely to the apparatus for treating statistically the observable atomic connexions. The representation shows how a physical system with a given statistical input evolves with time. It also discovers those statistical distributions that are preserved. But on account of the atomicity, not merely of matter but of all real connexions represented by the system, singular distributions in the sense of classical mechanics are physically impossible, that is representations of the classical type are inappropriate in the context of atomic mechanics.

Although the field theory seems to possess an apparatus for representing any microphysical system completely, it is not applied in this way. The fields are specified mathematically in relation to space and time or with respect to energy, momenta and other physical variables we associate with the field quanta—the particles. The latter show themselves in the world through physical changes in ordinary matter and through phenomena in which we can affect their imagined motion in the classical way. Since these processes are not represented in the theory, being regarded as unformalised details chosen by the experimenters using the theory, the field representation must be connected with the world on which we can act physically through the rules of interpretation for applying it.

The disembodied mathematical world of field theory supports the connexions supplied by the statistical theory in the limited regions of space associated with the interactions and resulting transformations of the entities that we regard classically as particles. This we do because we can observe their tracks when they are charged, or if neutral, because we can establish causal connexions between the events in which we imagine them to be produced, and with the other events in which we imagine the particles to interact with other things to produce observable effects. The theory establishes a system of connexions between the incoming radiation, the matter on which it is incident, and the outgoing radiations.

We are overawed by the success of the theory in practice and, finding that by its help we can establish agreement with experiment through quite long chains of reasoning, we are lured to believe that thereby we have proved that the apparatus of the theory represents nature in some occult way. That is, we want to go on and attribute to the mathematical representation used in the application of the theory a place like that which we know so well in the theory of our inventions. There we see what is represented; we select it, make it, and control it. But the mathematical apparatus we associate with microphysical representation does not work in this way. Its purpose is not to put something behind the phenomena but to establish coherent intelligible connexions between what we have observed in a great variety of observations and measurements. For we have no classical picture of what is going on in the continuum of the mathematical theory.

All of the form that the theory produces for comparison with experiment refers to statistical properties. That one uses particular microphysical forms for interaction and so on, merely shows how we proceed in the theory. This method of proceeding seems even today to have exhausted the possibilities open to it. In the theory of dispersion relations we see a new kind of reliance on mathematical ingenuity to provide new forms. These circumvent the difficulties in field theory associated with interacting fields interposed between the incoming particles and the outgoing products of interaction. Significantly, the virtual particle serves merely to determine singularities in the dependence of the scattering amplitude on the momenta of the participants in the reaction. Mathematical invention seems quite competent to establish the atomic connexion which serves to connect the values of momenta before and after the collision* and thus replaces the continuous connexion by classical force. So the physical interpretation in the ordinary sense is abandoned—even though the words 'causal' and so on are introduced in selecting the forms to be used. A decade from now the results derived from new experiments may show physicists a situation not unlike that revealed by the spectroscopists forty years ago. This will supply the new stimuli to theorists and guide them into new tracks of thought.

* G. F. Chew, *S-Matrix Theory of Strong Interactions* (New York: W. A. Benjamin, 1961).

CHAPTER XI

MATHEMATICAL INVENTION AND PHYSICAL REALITY

In a sense the investigation of elementary particles lies outside them. It should be contrasted with the ways in which the internal structure of living cells is revealed. Our physical experiments with elementary particles yield the information on which we interpret the individual instance of interaction between them by dealing with numbers of particles—preferably large numbers.

The processes by which we are led to new ideas are not usually represented in mathematical theory. This applies across the whole field of science. For example, modern knowledge of molecular biology depends on combining morphological and biochemical techniques for investigating how cells function: it has grown by a very complicated evolution through the work of men using a wide variety of techniques and a combination of superlative skill, imagination and inventiveness. The biological scientists have not arrived where they are by logical compulsion to accept the present view of things. They built up confidence in this view and in each other, so that the cooperative advance succeeds by continuing to reveal structure and function with ever-increasing detail and complexity of connexions.

The inclination of some philosophers to criticise what goes on here is like that of legal counsel defending a man indicted for a capital offence. Even in the court, however, the force of his argument, aimed at enlarging the area of doubt, depends in the long run on plausible reasoning. And so it is with the philosopher facing science.

The evolution of modern science is so intricate that a rewriting of it, to give effect to philosophical doubt by replacing current metaphysical commitments with new ones in an *a priori* manner, is possible only at the level of formal theory. Such commitments are analogous to giving up the Parallel Axiom to go from Euclidean to non-Euclidean geometry. In physics, however, theory has to meet not only the formal requirements of logical consistency in order to apply mathematics in elaborating its implications. It has also to serve effectively the men who need to use it in investigating physical nature, and in enlarging the scope of understanding man's physical environment, in the same sense as the modern biologist is deepening understanding of the structures and processes that sustain life.

It is good to institute comparisons between physics and biology. Mathematical theory is applied in biology to the physical and chemical processes going on in the structure revealed by the optical and electron microscopes and through the work of the biochemist. The theory plays a useful role in guiding thinking and giving precision to representation so that we may apply the sharp criteria of quantitative measurement in testing hypotheses. But the systems to which mathematics is applied have to be conceived. Even when mathematical theory is applied with the aid of a large computing machine, simple physical and chemical models must be constructed, and these models are chosen almost always in relation to the ideas that have been evolved in laboratory investigations.

Generalisation in biology—that is, the system of general ideas on which biologists are disposed to proceed in their thinking and rely on in their work—has not depended on mathematical abstractions like those that served to create and elaborate mechanics. The biologist has always been sensitive to the role of plausible reasoning in the evolution of his science, whereas the physicist, under the influence of the great mathematical masters, has been more ready to look on the logical compulsions of mathematical deduction as images of necessity in the real world. Formal systems seem out of place in a science that is evolving rapidly. Ideas serve a brief period to be replaced by others. Technical advance is so rapid that the experience of one decade is almost irrelevant to that of the next, that is, in a practical sense—like the early aeroplane compared with the modern jet or the early phonograph compared with modern sound-reproducing systems.

The point of these reflexions is to suggest that physics has much more in common with biology than some mathematical theorists or philosophers seem to think. The methods of physics have evolved separately from those of the biologist; they are based on different biases to man's experience. Nevertheless, the subject matter of study can be common to both biologist and physicist and may also engage the interest of the chemist, and only the methodology of science seems to separate what physics says about nature from the findings of the biologist. In thinking about the chemistry of metabolism, however, the biochemist is dealing with the physicist's atoms and the molecules composed of them. In oxidation and reduction, electrons may be transferred from one molecule to another. So long as we think about these things classically we are in no metaphysical difficulty. The electrons are part of the mechanism—little pieces of matter endowed with mass and electric charge. Together with the nuclei of atoms about which they execute some kind of motion, this is the stuff of atoms and molecules. But when we think of the entities of microphysics as the physicist has found them in quantum phenomena, the stuff seems to lose its permanence. Strangely enough, we seem not to be perplexed by light or other radiation in thinking about the existence of the molecules with which they interact. Notwithstanding that the energy required to support life is derived from solar radiation, and that in principle this energy contributes to the mass of the system which absorbs it, we do not think of it as an essential structural constituent: it is stored as energy of excitation or the kinetic energy of substantial entities.

The foregoing considerations bear on how we think about our physical environment—including our own bodies. If we think of our bodies as machines, as for many purposes we may, we have in mind the participation of the parts of the machine functioning cooperatively with each other in the processes of which its structure is capable. For this purpose, a part has a certain form and certain physical and chemical properties, and since in our representation we are not concerned with other processes that might change the part in question, the latter is regarded as not subject to essential damage in the processes that interest us. That is, our conception of permanence as a metaphysical absolute misleads us. We should consider it in relation to the processes that do not terminate the existence of form or transform substance. In practice, this is in effect what scientists and engineers do. Their representations are designed for particular purposes, so it is not necessary to represent possibilities in principle which are irrelevant for the purpose they have in mind. The wear of a machine is unavoidable, but if one is concerned with the performance of the machine over a short time during which the effects of wear would be negligible, no mention of wear appears in the representation. No process that could produce it is considered merely on that account. In spite of the obvious production of noise by large industrial machines, the mechanical engineer is not usually concerned to treat how the machine he designs is coupled acoustically to the ambient air. The physicist ignores the effects of X-rays in representing optical processes when he intends to apply his representation in practice to systems in which X-rays are not present in significant intensity to produce measurable effects.

Thus the entities of the physicist serve as substance for the biologist in so far as the latter is not concerned with processes that would invalidate treating them as substance. Where we get into metaphysical difficulty is in associating with the biologist's representation the possibilities revealed by physical methods that exhibit how our classical representations fail. When we think of the transfer of an electron from one molecule to another, we must regard it as an atomic jump, the electron being localised by electrical forces in one molecule and then localised at the other, but the timing of this process with infinite sharpness is prohibited and we are not allowed to represent the motion of the electron between the molecules with classical precision. We do not help ourselves by thinking of the electron as stored somehow or other in a wave passing between the molecules, and we must resign from specifying the electron's position within the molecule with classical exactitude. The electron in fact is the name given to the atomic connexion between the event at the first molecule and that at the second molecule. The situation changes as the result of the electron's passage just as drastically as the picture in successive frames of a cinematograph film of a very fast moving object.

Because of the existence of atoms of action, our representations of change at the microphysical level suffer the same kind of limitations that confront us whenever we have to place actual matter in space to mark positions. There are practical limits to the width of the division marks we inscribe on the scale, and every practical device we bring to our aid to improve resolution is limited in the fineness of interpolation it affords us.

Thus when we wish to think of nature supported by the existence of elementary particles interacting according to physical laws, as mathematical theorists are wont to invent in their formal systems, we are ignoring how this programme might be justified. Clearly the theory must explain not merely the possible combinations of these particles and the behaviour of these combinations, but this must accord with our experience of the properties and so on, they are intended to represent. It must also be consistent with the practical grounds for our classical metaphysical bias about the existence of atoms and molecules, and, indeed, of matter in general. Such a formal programme serves only as an exposé of the relations between our various methods of representation. It does not contribute to our confidence in these methods which we have learned to trust on other grounds. Rather it resembles schemes of organisation, and like them, contributes insignificantly to the activities it is supposed to govern, for these activities are engaged in and are carried forward because of human initiative motivated quite differently.

An electron by its mass and charge contributes to the substance of the atom or molecule of which it forms a part. Its existence is localised and endures with the atom or molecule just as an electron is imagined to be captured or lost by the oil drop in Millikan's experiment. But the electron in its transient passage between molecules depends on them, for its existence is an atomic connexion between them. Only in so far as its existence is signalled by an actual, or some imagined change in an intervening object consistent with other evidence, can we properly speak of the existence in space-time of the electron between the two molecules. The existence of the electron has to be supported by real physical connexions with objects much more massive than itself so that the limitations imposed by quantum mechanics concerning the specification of their position and motion are of little account in representing the electron.

The electron is thought of as a constituent part of an atom or molecule or a larger assembly composed of them. If by the action of light, for example, the electron is released, it ceases to be connected with the actual world until its energy and charge are localised, for example in changed substance of the more massive entity that captures it. In this way the existence of the electron has to be supported by objects that are regarded as enduring in the classical sense under the circumstances appropriate to our representation, for we ignore the thermal oscillation of the atom or molecule with respect to the much more massive pieces of matter that serve us in making physical measurements to localise them. By this way of looking at the matter we are trapped in a vicious circle. We require the concept of an electron to explain physically the properties of atoms and molecules, but in order to remove our metaphysical discomfort in facing atomic limitations on its properties, we are forced to rely on the existences of the entities with which it interacts, or at least, on the existence of the events which signal interaction by creating new existences that we can observe in the classical way as changes in objects we can see. This situation, however, should not surprise us. The existence of atoms and molecules is not conferred on them by the independent existences of their parts like the existence of the Solar System. They are in turn supported by the existence of ordinary objects, and our use of them to investigate the atomic and molecular structure of matter in particular situations in the real world. When we only imagine the molecules between which the electron is supposed transferred in chemical change, the question of their existence is irrelevant; nevertheless, if we wished to apply the picture we should know how to proceed, and our experience has established in us confidence that when we have practical reasons to expect the molecules to be present, we shall not be disappointed in our expectation. The actual process we devise, and depend on, may or may not use the electronic structure of the molecule. This does not matter. We are concerned with grounds for belief in the existence of the electron in the actual process. Our representation is based on the system of physical and chemical knowledge on which we rely in dealing with reality. That system is not concerned with a sort of Principia Mathematica of the logic of science; it is concerned with operating in this one universe we know, using some parts of it to investigate others, and thereby reach a better understanding of how it works. The idea that logical priority in theoretical exposition, as of logical elements in a system, corresponds necessarily to the order of substantial support for existence in reality, is just too naïve; it does not portray at all how men have developed the confidence in the ideas of theory on the basis of which they are willing to assert existence of the things and processes we speak of in science.

Nature's parts are interdependent: our theories reflect this in varying measure, but our formal approach to representation sometimes obstructs this. It is now becoming clear that in dealing with elementary particles—atomic entities of some kind or another—we are not to think of their possible existence *sui generis*. We have to regard them in much the same way as we must regard mathematical elements in a system of possibilities which have no properties until the formal laws that define the system are given. Because we can write down names for the entities a, b, c, ... and thus give the signs to which the formal laws of their combination will be applied, we appear to have the objects we have named existing in thought, whereas, until we have given the laws, they have no formal properties at all. In physics, of course, when we attribute to electrons, nuclei, mesons and so on, the masses and charges that have been found by experiment, we seem to have some substance to work on. But we are misled if we want to treat this substance as if we could associate it with space and time and mechanical laws in the classical way with any precision we care to prescribe.

In so far as modern physical theory represents the emission or absorption of a particle as a change in the field functional that keeps account of all possible interacting fields, it presupposes a persisting apparatus for observing, recognising, timing and locating the events. This apparatus is not represented in the theory, but in physical fact we depend on it to establish the connexions in our representation of the actual phenomena to which the theory is applied.

In so far as an electron can interact with other fields to create new particles we do not prescribe the mathematical place of the electron in our theory until the law of this interaction is specified. And of course, in complementary fashion we do not prescribe the mathematical place of the other particles until their laws of interaction are given. That is, we are forced to consider the system of fields as a logical whole. If we try to think of the fields as independent entities each of which we define by its own laws, and then superpose an interaction between them, we are using a method that depends on the weakness of the interaction to make it effective by approximation. It is like ignoring the reaction on the engine that produces a given force in mechanics or the effect of an electrical load on a generator.

These considerations take us into territory that is at present of much interest to theoretical physicists who have come to accept that in treating the 'strong interactions' between nucleons and mesons and hyperons, they cannot hope for the success that has attended the perturbation method in quantum electrodynamics. It is now recognised* that it is the interactions between the elementary particles that must give them their properties. At the same time, the methods of dispersion relations are being developed as the means of connecting ingoing particles with the emerging products of scattering or reaction.† These relations, of course, connect wave amplitudes and are not appropriate for representing on a microphysical scale what is going on in the locality where

* L. D. Landau, in *Theoretical Physics in the Twentieth Century*, p. 245, ed. by M. Fierz and V. F. Weisskopf (New York: Interscience Publishers, 1960).

† For philosophical interest, see *Dispersion Relations and Causal Descriptions*, by J. Hilgevoord (Amsterdam: North-Holland Publishing Co., 1960)

the reaction or scattering takes place. The connexions given by means of dispersion relations bear a certain resemblance to the engineering representation of the coupling of characteristic waves in a system in which multiple propagation is possible. These representations merely provide a scheme in which the numbers, that can be obtained in appropriate physical measurements, can be used to calculate the intensities and phases of the waves in the system under given conditions of excitation and loading. Likewise the forms of the S-matrix provided by principles of symmetry and unitarity must be supplemented by values of coupling parameters derived from experiment. The problem of the theoretical physicist must surely be to infer structure and function in the new context which his cooperation with the experiments should create. To argue that we have reached the end of investigation and that we cannot go farther is not very sensible.* If no road is possible by one method men will try another, and since the scope of theoretical innovation is coupled to the invention and use of the experimental devices and methods that can be applied effectively,† theorists must look on their work as effective in proportion to the extent that it opens the door to fresh investigation, not in getting rid of unsolved problems. In this regard, the influence of experimental science, especially in biology, should diminish the influence of the academic penchant for closed formalism, and keep before us that, forms of connexion and organisation other than those that have traditionally interested the physicist and engineer are exemplified in the functioning of organisms. We should be disposed to expect that the study of organisms may bring into physics new

* The essential relation of theory to experiment is not merely that of a musical performance to a score—a mere comparison and the judgment 'correct' or 'not'. In the past, appeal to experiment challenged men's thinking by laying down in practice a method that makes growth in knowledge possible because of feedback; experiment stimulates exploring ideas and this in turn suggests new experiments. The objective professed for an investigation when it starts may fade away as knowledge and understanding grow. Having in mind to produce something final and definitive applies only to the limits known now as to what may become possible. One has to finish an activity only because of practical considerations arising from the environment of other human interests and circumstances that affect its continuance.

† In this connexion, the discovery of the Mössbauer effect is pertinent.

intellectual tools for investigating inanimate matter. Already, study of the brain and nervous system has shaken confidence in the formal attitudes associated with the logic of computing machines.*

At the same time, these considerations should arouse some scepticism towards the idea that the physicist's problem lies 'in the elaboration of a code, by means of which a one-to-one correspondence between observations and a certain mathematical scheme may be established',† For this approach with its appearance of logical regularity ignores the circumstances of significant use which is not a matter of mathematics at all.

One of the justifications offered for formalism in science is that it eliminates anthropomorphic bias in our attitude to nature. One might legitimately ask 'what of the bias that introduces the particular mathematical apparatus we depend on?'. There is the bias of continuity mathematically defined, as opposed to vague ideas; the bias of covariance and so on. To make progress we cannot escape committing ourselves to something, but let us admit what we have done. Such a commitment was made in physical theory when it was fashionable to assume the conservation of parity in all interactions between elementary particles. It was not questioned until Yang and Lee recognised that in order to understand the experimentally observed disintegration of K-mesons into two or three pions, this conservation theorem must be violated, because if one of these modes of disintegration conforms with conserved parity, the other cannot. The philosophical atmosphere associated with this achievement is accounted for in the following way. Yang and Lee changed the attitude of physicists to their basic procedures in representing interactions of elementary particles and nuclear processes. They exhibited as open to question something that was taken for granted.

In philosophising about physical theory we must consider not only the logic of representations and how we use them, we must

---

* See E. Nagel and J. R. Newman, *Godel's Proof*, pp. 100 *et seq.* (London: Routledge and Kegan Paul, Ltd, 1958); also J. von Neumann, *The Computer and the Brain* (New Haven: Yale University Press, 1958); W. Ritchie Russell, *Brain, Memory, Learning* (Oxford: Clarendon Press, 1959).

† F. Villars, in *Theoretical Physics in the Twentieth Century*.

also reflect on the ways of resisting, accepting and promoting particular methods. Here the attitudes of Blackett* and Pauli† to non-conservation of parity should serve to enlighten us. Physics is not a legal system with authority based on legislative action, books, and legal arguments. Certain technical procedures of experimental physics, it is true, are subject to national laws and international agreements, but even these are matters for discussion and may be changed. There is indeed a polemical area in which attitudes of mind affect theoretical invention and writing in physics. These attitudes are inescapable concomitants of men living in the world and engaged with each other in a wide range of activities. To pretend that such differences can be excluded from science, or that it is characteristic of science that they should be, is inconsistent with the history of its growth.

The process of constructing theories has in it an element of finding physical significance for successful phenomenological descriptions. It is exemplified in the interpretation of atomic spectra, the history of which has been impressively exposed by Kronig in the *Memorial Volume to Pauli*.‡ More often than not when real novelty appears, what happens is that the experimental results are organised in such a way that some mathematical form is exposed and recognised. Then it becomes the goal of theory to invent an explanation for this particular form, either exactly, or as a good approximation. The criterion for accepting such explanations is that the theory is open to other tests.

It is surely remarkable that the possibility of admitting complex values for dynamical variables that suggests itself in the elementary consideration of the wave equation for a free particle should find an echo in the subtleties of the dispersion relations of modern quantum field theory.§ In these relations we have some characteristic of physical interaction such as scattering amplitude which is a function of the energy, and we want to continue the functional

* P. M. S. Blackett, *Physics Today*, vol. 14, no. 2, pp. 86–7 (February 1961).
† C. S. Wu in *Theoretical Physics in the Twentieth Century*, p. 272.
‡ Op. cit. *The Turning Point*, pp. 5–37.
§ See Bogoliubov and Shirkov, *Introduction to the Theory of Quantised Fields*, ch. IX, pp. 558 *et seq.*

relation into the upper half-plane of the complex variable E. By Fourier transformation this process leads us to consider a function of the time, and on this second function a 'condition of causality' is imposed, namely it vanishes for $t < 0$. What is done here is to exploit a complete system of mathematical possibilities for representing physical phenomena. This same consideration is relevant to much of the mathematical involvement of both quantum mechanics and field theory, and it is well to note that mathematical generality or some other attractive aspect of form may mislead us with respect to physics. There is certainly no necessity that mathematical inventions must succeed when they are applied; it is men who treat nature in such a way as to render the mathematical method rewarding. The integrity of the formalism provides an element of practical advantage in the context of its use, and economy in learning to understand it and apply it fluently. Integrity is a matter of internal logical structure, whereas the force of the formalism in physics is determined by external relations by which the formalism is applied.

Are we to regard theoretical physics as the evolver of techniques for deriving formulations that can be compared with experiment, or as our teacher as to how we should imagine nature? If it is the latter, it has an obligation to alter ordinary language—think of the round Earth, the Copernican system, and Newton's law of gravitation as examples of the currency achieved by effective new concepts—but it cannot do this without metaphysical insight. Something as effective as the spherical map of the Earth is needed. Classical physics provides this in respect to crude descriptions, but it does not really help in explaining the essential conceptual difficulties presented by quantum phenomena because we need new ideas. To say that we *must* depend on highly abstract mathematics to deal with atomic representation may be gratifying to mathematicians, but it will not be accepted by men who appreciate how new ideas in science and mathematics work their way into the fabric of our civilisation in a very great variety of ways. The formal nicety of the academic man is largely irrelevant in this process which depends on imaginative teaching. We have to make inventions to instruct.* In spite of the dictum of Rosenfeld alluded to early in this book, explaining the implications of quantum physics for our concept of atoms and the logic of conceiving them is still incomplete.

We do not touch physical reality with mathematics, but we are involved with reality in living, in the activities of which we speak through ordinary language, and in particular in the activities that engage the experimental scientist in his work.

A system of ideas is scientifically germane to living only when we use them to govern our activities in studying nature. That we use one system rather than another is decided by experience—not only of one man, but of many men, not just at one time, but over generations. Experience reveals when and to what extent ideas are appropriate, and confers on us the confidence to commit ourselves to them. In this mood we are sure of the reality of the world seen through the whole system we have learned.

Through mathematical invention, however, men have explored other systems and made clear that there are available to us many systems that we do not use in science, although they are just as valid logically as those we do use. Accordingly, mathematics seems continually to present lures for thinking about the world in terms different from those to which we have grown accustomed. It is quite natural that the mathematical inventor should wish to see aspects of phenomena in terms of the forms he has created. Sometimes his view achieves a measure of acceptance by his colleagues, sometimes it does not. If his view is fruitful in interpreting experimental facts, it gradually wins its way into the language of physics and is relied on as other ideas are that men use in scientific investigations. So it comes about that successful mathematical invention begets the disposition on the part of some men to regard mathematical activities as leading to deeper understanding of reality, and to view the compulsions imposed on the mathematician when overcoming technical difficulties in his work as if they were

* Cf. what has been done with planetaria to exhibit discoveries in astronomy, the models to explain nuclear fission, and the modern successes in explaining mathematical ideas to a wide public through skilfully chosen illustrations, models, and other concrete aids to understanding.

images of the reality he represents. The history of mathematics, however, shows that the difficulties of one generation may very well be overcome in the next by fresh invention. The idea of mathematical systems defined without reference to their possible use and not subject to evolution as the result of experience in using them seems a formally attractive one. Nevertheless, mathematical ideas do change and old systems are replaced by new ones. Accordingly, the resources for operating with reality are not static even in respect to logical structure; speculation invites fresh experiment and investigation. On the other hand, experience may convince us that the ideas with which we approached phenomena do not succeed and impel us to look for new ones. In this process mathematical invention may have a part to play in contributing to the formal elaboration of the plausible reasoning of the scientist. Eventually the formal system dominates thinking about reality, and the language of science acquires new resources.

The more sophisticated the mathematical invention, the fewer the men who understand it and can use it effectively, and the less real influence it has in the thinking of men about nature. Are we then to say that for this cause, appreciation of the discoveries of physical science—and indeed of other science in which mathematical reasoning plays an essential role in its growth—must inevitably be restricted to the few? Faced with the corresponding practical problem in engineering, inventive men have devised ways of overcoming it. Instruction in theoretical ideas is matched to the training, capacity and activities of the men who are to use them. Likewise in science, in so far as mathematical invention is important, the conceptual implications of it have to be explained in a variety of ways. The mathematician's approach to this matter is more likely than not to over-emphasise the logical integrity of the system, the engineer's attitude will over-emphasise the effectiveness and economy of the system for computing. To the physicist the important elements relate to understanding phenomena and going on to fresh investigation of reality by experiment and by observing in new ways natural processes that go on beyond man's control. For in an important sense science is part of man's adaapttion to the world in which he lives. Getting accustomed to ideas of it and learning new ones from his experience are the intellectual core of his activities. In the process he may learn something about himself and free himself from errors in thinking that descended to him or that he has only too readily committed himself to as a social animal.

We depend on mathematical invention for our ideas of space—the basis of all representation—for it deals with systems of logical possibilities and connexions between them. In so far as these inventions are found appropriate and useful, we learn to rely on them, and, indeed, adopt their logic by means of the axioms in our formal theories in science. The axioms seem reasonable, even self-evident to the man with a particular metaphysical bias. He is ready to accept the logic of the theory as an intuitively satisfactory way of representing nature. Nevertheless, this is irrelevant in justifying the theory as part of science, except in the following respect. When theoretical ideas are congenial to men, they are more likely to apply them with confidence and to commit themselves to new invention in using them.

The history of the development of quantum mechanics exhibits how a mathematical system can be elaborated through the successive inventions that elucidate the forms of phenomena revealed by experiment. In a minor way this process goes on all the time in physics. What is especially significant about quantum mechanics is its scope at the foundations of physical representation. It does not use the kinds of space to which we were accustomed before its invention. The success of its methods, established with overwhelming weight of evidence, surely should lure us to try to understand what these methods show us about our basic logical approach in representing atomic phenomena and, in the process, to evolve a concept of atoms that is logically integrated with it.

Our attitude to physical space is based on men's experience with the ordinary objects that can be moved about; it is inextricably interwoven with metaphysical assumptions about the persistence of the objects and continuous motion. Ordinary space is a continuum with only three dimensions, and time has one. The space of atomic existence, however, has infinitely many dimensions, one corresponding to each point of physical space. We therefore have to look at physical space as a set of disconnected logical possibilities, everywhere dense, but not forming a continuum. Between pairs of these possibilities, transformations which correspond to displacements in ordinary space do not constitute a continuous family, they are atomic connexions. Two points of physical space, which on the continuum view we regard as infinitely close to each other, are still distinct in the sense that the x-axis is distinct from the y-axis of Cartesian coordinates. No matter how we interpolate arithmetically, we never by the subdivision of an interval on physical space approach the logical possibility represented by a point of physical space lying between the ends of the interval on the line joining them; for it is an independent logical possibility, whereas the interpolation is tied to the ends of the interval.

Thus the logical possibilities must be symbolised in such a way that this is made evident. The singular distribution $\delta(x-x_1)$ is attached to the point x_1 of the classical continuum of x; to each point x_1 there corresponds one distribution. These distributions are the basis vectors of function space. They are independent possibilities in that we cannot express one as a linear combination (superposition) of others. If we pass from x_1 to x_2 in physical space we represent the transition as a unitary transformation of function space analogous to rotation of coordinates in ordinary space. The continuity of the physical space of classical physics is reflected in the continuous group of unitary transformations by which are reached the continuous series of distributions $\displaystyle\int_{-\infty}^{\infty} f(\xi)\,\delta(x-\xi)\,d\xi$ through the continuity of the function $f(\xi)$. We pass from the singular distribution $\delta(x-x_1)$ to the singular distribution $\delta(x-x_2)$ through distributions over the whole range of x. In this way we recover the advantages of continuity for mathematical analysis. If because of physical laws to which $f(\xi)$ is subject, it is not possible to pass through the singular distribution $\delta(\xi-x')$, where $x_1 < x' < x_2$, there is in this transformation no point of physical space between x_1 and x_2 for the process in question. The most elementary example of this is the atomic connexion between x_1 and x_2.

These considerations are relevant to thinking about the relation of the lattice to the continuum. In the above sense, the step from one point to an adjacent point of the lattice is an atomic step on the corresponding continuum, but we may regard it as a continuous transition in *another* space. The idea of the atomic connexion, step, or transition, as represented outside the space in which we normally represent the two possibilities being connected, is relevant to our thinking even about atoms of quantity of substance, for we ordinarily think of quantity as continuously variable (we have already referred to the practical reasons for this). We do not ordinarily think of the atoms of electric charge, for instance, as requiring us to change our attitude to the space in which we represent the magnitude of electric charges; we apply the continuous representation with the rule $q = ne$, and do not regard the change of charge as a continuous process—an atom of charge is imported or removed by an electron, or other charged atom. But the atom of action is connected with motion in physical space; we cannot regard it merely as a substance tied in atomic packets. How we should think about it is to be found in the forms of the successful mathematical inventions for representing atomic phenomena. In looking, we must try to dissociate the devices that aid the mathematician in managing the representation and the calculations by means of it, from the essential elements that are relevant to determining a new metaphysical grasp of microphysics. These are the elements that are essential for representing atoms and their motion. Planck's constant merely determines the scale of the phenomena for which the quantum effects are evident in mechanics.

Basically the mathematical apparatus for connecting the continuum and discrete spectra of the possible values of physical variables (of which lattices are the simplest examples) was exhibited in Weyl's *Group Theory and Quantum Mechanics* over thirty years ago. In it (p. 275) Weyl speaks of having found 'a very natural interpretation of quantum kinematics as described by the commutation rules. The kinematical structure of a physical system is expressed by an irreducible Abelian group of unitary ray rotations in system space. The real elements of this group are the physical quantities of the system; etc.' That is, he was satisfied to reveal the place of the algorithms of quantum mechanics in mathematics. This kind of interpretation has always played a very important part in the history of theoretical physics. Indeed, the invention of quantum mechanics as a system for dealing with atomic physics consisted really of a succession of steps by which it was revealed in this way.* But this is not the kind of interpretation we seek here. We accept the mathematical elucidation together with the logical system it offers for physical representation. We are concerned with understanding how the success of these mathematical methods reflects on our predicament in the face of atomic phenomena. We shall not find the answer in recounting the historical details of how the mathematical system was recognised, and how the proper ways to apply it in physics were discovered. We have to make simple models that we could present by physical apparatus and engineer them so that they show us observable processes in space and time of the kind that would be effectively represented by the methods of quantum mechanics. The model is required, not as a model to explain atomic behaviour, but as a means of explaining how our accepted method of representation is relevant to atomic phenomena. In the process we learn a new meaning for 'atomic', in that its scope extends much farther than the simple concept of an atom of chemical substance that we have been able to cope with in the classical way, viz. the atom is represented by a point in space, it endures in time, its mass and other scalar properties are the atomic steps between consecutive values of the corresponding physical variables we use to describe the properties of a piece of the substance. The essential logical element in the idea of atomicity is that an atomic connexion between elements of the space of a physical variable cannot be represented as a connexion in that space. Discontinuity and unity are relevant to it.

* See B. L. van der Waerden in *Theoretical Physics in the Twentieth Century*, pp. 199–248, ed. M. Fierz and V. F. Weisskopf (New York: Interscience Publishers Inc., 1960).

In one of his penetrating remarks, Pauli said* 'It should be required that atomicity, in itself so simple and basic, should also be interpreted in a simple elementary manner by theory and should not, so to speak, appear as a trick'. This metaphysical reflexion by a master of mathematical invention in physics serves to remind us that in confronting physical reality, men are the active choosers of the forms by which they represent it and try to understand it. If we wish to understand why a particular choice is made we have to learn to understand men and how they are committed to life.

* *Pauli Memorial Volume*, p. 189.

CHAPTER XII

A VIEW OF ATOMICITY

When we approach atomic phenomena with the ordinary strong bias to classical ways of thinking, we insist on motion as continuous motion, existence as continued existence in ordinary space-time, substance located in the space of our representation, and causal connexions between events specified with the refinement of continuous representation. This way of picturing nature has occupied so eminent a place in the thinking of men since science began that it is very difficult indeed to free them from the compulsion to follow it, because they are so ready to conform. How can they be led to appreciate that there is a limit to this method of representation, that continuity is our invention, and that experience in quantum physics shows us that we have no classical explanation of atoms for the method will not work in microphysics? Provided that we are not involved where atomicity has full sway, we can use the classical metaphysical bias to good purpose. Physics is not concerned with upsetting such effective attitudes but it has revealed where they are not appropriate. In this book we have been concerned with understanding how the existence of atoms sets a natural limit to the successful use of the old method. If one feels that thereby one's world is turned upside-down, it is time to recall the basis of our confidence in scientific method. It is found not in our formal theoretical inventions, but in the naïve realism of our learning the methods of science to investigate nature, and leaning on the teaching of experience about phenomena and how we may interpret them. That we have to learn a new method because of the technical successes in atomic, nuclear, and elementary particle physics should not surprise us. It is no more remarkable than the new practices to which we must accustom ourselves through understanding new ideas in the world of modern engineering.

Our simple idea of an atom is an atom of substance: the quantity of substance is not continuously variable. Just as cents are the least denomination of decimal currency, and so one cent is the least difference between sums of money that can be transferred in actual payments, the atoms of chemical elements are regarded as units in some physical transactions. However, since many elements have isotopes with differing masses, the idea of atomic mass for these elements is applicable only as an average; so the determining quantity is of course the nuclear charge of the atom.

In a game played on a board, we meet atomicity in a different way. The individual moves on the board are discrete. A particular move on the board is an atomic connexion. It is a transformation of the arrangement of the pieces, and it is named by giving the initial and final position of the piece moved. In a continuum, an atomic step or jump is an integral connexion between points: it has no parts that connect intervening points.

When similar projectiles are shot with the same muzzle velocity at different elevations, they have different ranges on the ground; that is, each projectile makes an atomic step or jump on the ground. There is no intervening point on the ground between gun and target at which the projectile is found. Yet we know that it has a continuous path through the air, and we compute where it will fall given the angle of elevation at which it is fired. We are not impressed with the fact that we cannot find the shot on the ground because we can find it in the air. But this is not how the phenomenon is presented in sound-ranging. There the sound of the gun is timed and located and the explosion of the shell is timed and located. What happens in between can be inferred by dynamics, provided that there is no doubt that the arrival of the shell is to be connected with a particular firing event. Nevertheless, sound-ranging yields no observation of the intervening passage.

As we deprive ourselves of the means of interpolating observations between the initial event and the final event to which we connect it in thought, we tend towards the atomic connexion between the events when there is in principle no interpolation possible at all.

Because the events are at different places at different times, the step from the earlier to the later is associated in our minds with motion. If mass is transferred in the atomic step, we associate with the step a position in momentum space. We must not expect this position to be sharply defined for we are unable to apply the limiting definition of velocity at a point as the derivative of displacement with respect to time. Nevertheless, through the momentum the atomic step is associated with the change of the value of the mechanical action.

In classical mechanics the momentum field may be derived from that of the action (S) by computing its gradient in x-space. Thus the atomic connexion which we have in mind is a step in x and in S, associated with a value of the gradient of S with respect to x. Since the uncertainty in the value of the momentum (p) multiplied by the range in x is substantially equal to the uncertainty in the change of action, and since the latter is quantised in atoms of magnitude h, the product of the ranges in x- and p-spaces associated with the atomic jump cannot be less than h. Here we exhibit the uncertainty relation as connecting the uncertainty in the derivative of a function, the relevant range of argument of the function, and the atomic step in which the function is quantised.

Let us consider an example that takes us away from the associations of quantum theory in order to make this point clear. Consider the collection of a 10 cent bridge toll. Imagine that over a long period of time the rate of collection is constant. Our expectation as to the cash received in time t is given by $C = R\,t/10$ dollars. Clearly we do not apply the mathematical equation $R = 10\,dC/dt$ to compute our expectation of the rate at which cash is accumulating over a short interval of time. The least time interval that is significant in relation to the accumulation is the reciprocal of R. Likewise the product of the time t and of the uncertainty in R over it cannot be less than unity.

We may consider a different illustration. In making topographical models it is common practice to build the model in layers of equal thickness cut to the shape of the contour line at the level at which the layer is to be laid. The layers are properly centred and oriented in the pile to make the model. Now think of the corrugated surface of a hill in the model. The gradient of its slope is the rate of change of height with respect to horizontal displacement. In the model, height is quantised in steps equal to the thickness of the layers. We cannot apply the differential calculus to the model to find the gradient of the smooth hill which the model represents, for the model does not present the infinite possibility of interpolation consistent with our idea of the smoothness of the hill. Thinking of the application of the model and being satisfied with finite differences, we ignore this limitation, and we regard our representation as approximate. Of course, in actuality the surface of the Earth need not be smooth, and it rarely shows the precise layered structure we have imagined. If it did we should be confronted with a barrier to the continuous representation to which we are accustomed. Quantised height sets a limit to the precision of the gradient and the position at which we may attach the gradient on the map.

Here we have looked on atomicity of a function (height) of a variable (position on the base horizontal plane) as limiting the precision of defining the gradient of the function. Our classical bias tempts us to invent interpolated values of the function and to regard the actuality as incomplete, for we have in the past always regarded limitation on interpolation as merely practical due to lack of refinement in measurement. What is more, we are quite sure that the function should be represented as changing smoothly without steps.

In order to preserve the effect of our continuous picture when crudely applied, and yet acknowledge the atomic structure of actuality revealed to us by observation and reflexion, we require a mathematical invention.*

* We introduce a continuous function $\psi(x, y)$ of x and y the coordinates on the plane of the map. The gradient g_x in the x-direction is represented by means of the equation

$$\frac{\partial}{\partial x}\psi(x, y) = \frac{2\pi i}{h} g_x \psi(x, y),$$

where h is the vertical step of quantised height, $i = \sqrt{-1}$. We may proceed in analogy with quantum mechanics in its dependence on classical forms to set up the wave equation for ψ. The classical form plays the role of our concept of the geometry of the smooth hill. This illustration is a suggestive one as the reader

Our idea of the implications of 'atomic' for the language of physics has to be broadened to remove our philosophical difficulties. It is not enough to bundle substances into packages of minimum content. We must disengage our thinking from many of our habits associated with continuity. The existence of atoms shows us that the logic of ordinary continuous representation loses contact with the real world of physics. So the forms of speaking about the world using that logic confuse us when we try to imagine what is going on in an atomic phenomenon. These forms apply to the ordinary things and processes that can be represented classically. Existence of things in the world is understood in this context, but an atomic entity is not—that is, in so far as it is an atomic connexion in the world connecting events, and of course connecting places in the spaces of the physical variables we use to describe the changes that support the existence. We speak of the atomic connexion of events because this is the simplest; but we may also connect other elements of classical representation, for instance the beam in an accelerator and the instrument in which we detect the radiation.

We associate substance with an atomic connexion—something transferred in the jump. For example, an electron transfers its energy, momentum, spin angular momentum and electric charge. So the event A initiating the connexion is associated with a transformation in the space of these quantities corresponding to the departure of the electron, and correspondingly the inverse transformation is associated with the event in which the electron is captured, or better, with the event B that terminates the atomic connexion AB. When we choose to think of the atomic connexion as the motion of the electron, we want to imagine the mechanical properties mentioned above transferred as properties of the motion. The space in which the motion is imagined is continuous; we are thus lured easily to conceive of the motion as continuous like that of a classical particle. Once committed to this view of the

will find if he studies it. We shall only note here that there is not much point in speaking of the slope in our model where the classical hill has great curvature, for instance in a hillock on an otherwise smooth slope.

matter we have lost contact with the essence of atomicity in respect to the apparent motions treated by quantum physics.

The individual connexion AB connects only the event A with event B. It is related to motion only as a member of a family in which A and B are variable in a field of events. The motion is represented by a distribution of amplitude and phase over the field of events, but it is not the motion of a particle. Just as the connexion of particular events at different times leads us to think of it as physical motion of a particle, so the connexion by the law governing the distribution over the field of events leads us to think of it as a physical wave. Since the space in which we imagine the wave to be propagated is continuous, we are easily committed to the classical idea of a wave process, and accordingly lose all contact with the atomic connexions constituting the physical reality we can observe.

In extricating ourselves from philosophical discussion of atomicity and motion let us not forget that we have our pictures of atoms in matter revealed by X-rays, neutrons and electrons, by the field ion microscope and by the Moiré patterns obtained by electron diffraction to show dislocations in a crystal. Because we can present these pictures showing objects in space, they are pictures in the classical sense for certain uses, but they are not used in the classical way in physics for they are applied and indeed have been produced statistically. We could not have the pictures and interpret them as we do if the process we use to 'take' them did not provide means of adequate resolving power through the statistical properties of a large number of quanta or atoms used to study the structure shown in the picture. On the other hand, we cannot have this kind of picture of electrons in an atom or nucleons in a nucleus, because inside the atom or nucleus they are not localised sharply enough.

So far as classical representation is concerned, a neutral atom of hydrogen, for instance, consists of a proton and electron. The latter is localised in the vicinity of the former. The atom has no net electric charge because it is not acted on by an applied static electric field that we can easily produce in the laboratory.

We think of the atom as a permanent entity in a certain atomic state until through interaction with another atom or other physical system it changes that state—a photon may be emitted or absorbed, the electron may be ejected. The emission of the photon connects the two atomic states. The ejection of the electron connects the ion with the neutral atom. If we wish to probe farther we must examine the particular physical circumstances in which the change takes place, so that eventually the ion and the atom are related to our apparatus for detecting them. On occasion, we may be concerned with statistical conditions—a beam of atoms and a beam of ions. These are essentially classical agents. In these circumstances we have in fact to do with atomic connexions between the classical agents.

We have no trouble about the crude localisation of the electron in an atom, but we wonder what is going on in the atom between the significant physical events in which the atom interacts with its environment. This resembles the problem of connecting an emission event with a detection event in a beam of radiation. We want to think of motion through physical space: interference shows that we cannot have this. With the atom, we are interested in connecting events at substantially the same place—we are not interested in the possible motion of the centre of mass of the atom—but at different times. The persistence of the electron in the atom is not represented by a real process in time but by fixing its energy state. With this state is associated a certain statistical distribution that plays a part in the connexion between different times like the intensity of radiation, but only if it is used in a particular way, with a certain crudeness in the resolution of time.

Our quantum atomic models should be considered in relation to their use in physics. To speak of the atom of hydrogen through the classical analogue is a convenient way of excluding philosophical questions from the physical discussion; it permits one to concentrate on the calculations by means of the proper model. By the theory of measurements the latter suppresses questions about what is going on. At the same time it seems to represent the existence of something—the wave function which depends on space and time.

One speaks of the electron cloud about the nucleus of the atom. It endows the atom with size and, as we have already remarked, it may be used to interpret certain phenomena in a simple way. Surely it is clear that our picture of an atom is a statistical picture; its use is statistical and we are driven to this because we have to deal with atoms both of substance and of connexion. The only way for us to have a picture in continuous space-time is through statistical interpretation. The picture does not inform us about the individual atomic event or process, only about the probability of certain happenings and the expectation values of physical quantities we can measure.

What is remarkable is that the Rutherford–Bohr picture transcribed into wave mechanics works as well as it does. The mathematical invention seems to pass beyond the classical picture we think we understand, but at one stroke it deprives us of all the supports of our thinking about the physical objects we imagined in the classical way—continuous motion in continuous space and time, with causal connexion between events and so on, specified with limiting precision.

The spatial picture of a chemical atom is required neither by the atomic nature of its parts nor by the existence of Planck's constant, but as a convenience to represent the atom in the continuum we use in physics. We have learned to compute energy levels and a great variety of significant atomic magnitudes with the aid of the more elaborate mathematical models derived from the early one. These computations help to carry physics along as a technique with manifold applications.

The modern model of the structure of RNA or DNA molecules, derived by the methods of biochemistry and biophysics, exhibits chemical structure by the arrangement of the atoms in space. The precision of the model does not imply the exact fixing of the positions of the atoms. It shows the classical object that corresponds in certain important statistical respects with experience of the substance that is separated in the ultracentrifuge, undergoes biochemical transformation and participates in genetic processes. The classical motion of the parts of the model, however, does not represent the atomic connexions appropriate to modern quantum chemistry by which the physics and chemistry of the molecule are related to the physics of atoms, electrons, light and so on. The structure of the molecule shown by the model may be likened to the simple picture we have of a large electrical machine. Without the physical means to detect the mechanical vibrations and electrical processes going on, it appears static. Our only clue to its activity is the hum we hear. We are in no doubt as to its existence, and though we may be ignorant of the technical details of representation necessary to the engineer who uses it, we are ready to accept that he knows what he is doing. May we not adopt a similar attitude to the scientist dealing with the atomic structure of matter and radiation and the molecular structure of biochemical substance? Of course we may, and often do. It is when scientific discoveries seem to challenge formerly accepted ways of thinking about nature and ourselves that we are lured to philosophise as if by logic or rhetoric we could adapt ourselves to the new, either to accept or reject it. Without the experience of the scientist it is usually hard to accept the compulsion he feels in his thinking when he has acquired successful inventions in his work. Although faced with 'the conceptual difficulties' presented by quantum mechanics, physicists have learned to go on with confidence in its methods and with great success have developed new inventions related to them. Nevertheless, they have not yet cleared the ground of understanding how we should speak through ordinary language of atomic reality and the logic of atomicity.

In the process of clearing the ground, we find philosophical difficulties through questions like the following. 'How can memory have a physiological or biochemical mechanism unless the protein molecules imagined to carry the information endure continuously?' (This question is based on recent experimental evidence that points to protein molecules as storing information about experience in the nervous system.) Here classical metaphysical bias misleads us. We want to think of a mechanism that will carry something from the past into the future. In continuous representation, atomic connexions seem to leave a gap in which the information would be lost or distorted. Couldn't just that happen to the protein molecule imagined as existing continuously in its complicated environment? The function of the atomic connexion is to bridge the gap; it connects the events in which physical effects occur, just as the classical picture of the existence of the protein molecule. In using the classical picture we depend on habit; for we have no ordinary experience of interpolating observations of the things we see on a time-scale more refined than a few milliseconds, and even for physicists there is a practical limit to refinement in timing. So we have no logical support for the view that objects *must* exist continuously. Indeed, that is not what we need. We want the sure connexion of an experience now with a biologically effective memory of it in the future. We do not have a sure connexion classically; the molecules may suffer chemical change that destroys the record. Thus we think of a statistical distribution of the classical connexions, just as we are compelled to do in treating atomic connexions.

The sickness of a time is cured by an alteration in the mode of life of human beings, and it was possible for the sickness of philosophical problems to get cured only through a changed mode of thought and of life, not through a medicine invented by an individual.

LUDWIG WITTGENSTEIN,
Remarks on the Foundations of Mathematics,
translated by G. E. M. Anscombe

INDEX

accelerator, 153, 159, 209
accounting, 30
-operator, 150, 168 f.
-system, 180
action, 55, 67, 110, 168, 190, 202, 207
Akhiezer and Berestetsky, 116, 142,
171
angular momentum, 150
annihilation operator, 151
antenna, 54, 121, 151
anti-particle, 173
Ashby, W. Ross, 43
astronomy, 86, 128, 198
atom, xii, 87, 132, 159, 200, 203, 205,
atomic connexion(s), xi 70, 100,
166–7, 202, 206
family of, 132–3
atomic entity, 59, 105
atomic existence, ix, 90 f., 165
space of, 200
atomic theory, 33
atomicity, xii, 121, 125, 128, 130–1,
183, 203, 205 f.
Augustine, 35, 155

β-decay, 29
biology, 187
Blackett, P. M. S., 196
Bogoliubov and Shirkov, 176, 196
Bohm, D., 8, 50
Bohr, N., 4, 5, 128, 212
Boltzmann, 42
boson, 142, 171
Bridgman, P. W., 8, 31
Brownian motion, 59, 78, 88
bubble chamber, 14
bullet, 93, 153

calculation, xi, 24, 37
Chadwick, Sir James, 16
chemical detector, 109
chess move, 91, 128
Chew, G. F., 185
Clerk Maxwell, 1, 147–8
coherence, 49, 140, 142, 165
commutation rule, 142, 170, 173, 178

complementarity, 5, 56
Compton effect, 157, 159
computer, viii, 10, 24, 37, 127, 187,
195
confidence, 10, 24, 36, 41, 73, 86, 89,
102, 163, 186, 192, 195, 198, 200,
213
continua of classical physics, xii, 63, 144
continuum, 95, 110, 202
'Copenhagen school', 29
cosmic rays, 16
Coulomb field, 151
counting, 117, 145
coupling of fields, 152
creation operator, 150, 171
process, 172
cross-section, 154
crystal, 21, 108, 159

de Broglie, L., 6, 7, 8
Dee, P. I., 16
'deep' questions, 3
derivative, 124, 125
Descartes, 156
detector (of interference), 49, 60
diagrams, 26
differential calculus, 6, 66, 123
diffraction, 13, 53, 61, 90, 100, 109,
145, 161
Dirac, P. A. M., 24, 28, 60, 123, 129,
130, 151, 164
dispersion relations, 185, 193, 196
distribution, 112, 124, 139

Eddington, A. S., 42
Einstein, 67, 76
electromagnetic field, 149, 169, 182
electron, 87, 150, 155, 158, 188–90
electron-positron pairs, 156–9, 183
elementary particle, vii, 4, 56, 150 f.,
186, 192, 193
emulsion photograph, 14, 87
energy, conservation of, 149, 152
engineering, 24, 33, 35, 149, 199, 205
Estermann and Stern experiment, 108
Euclidean space, 57, 147